CHOOSING SELECTION

CHOOSING
SELECTION

· · · · · · ·

The Revival of Natural Selection in
Anglo-American Evolutionary Biology,
1930–1970

· · · · · · ·

STEPHEN G. BRUSH

AMERICAN PHILOSOPHICAL SOCIETY
Philadelphia · 2009

TRANSACTIONS of the AMERICAN PHILOSOPHICAL SOCIETY
Held at Philadelphia for Promoting Useful Knowledge
Volume 99, Part 3

Text and cover design by E. H. Graben.
Set in Adobe Caslon with Waters Titling display
by Graphic Composition, Inc., Bogart, Georgia.
Printed and bound in the United States of America.

ISBN-13: 978-1-60618-993-1
US ISSN: 0065-9746

LIBRARY OF CONGRESS CATALOGING-IN-PUBLICATION DATA
Brush, Stephen G.
Choosing selection : the revival of natural selection in Anglo-American
evolutionary biology, 1930–1970 / Stephen G. Brush.
    p.   cm. — (Transactions of the American Philosophical Society held at Philadelphia
for promoting useful knowledge, ISSN 0065-9746 ; v. 99, pt. 3)
Includes bibliographical references and index.
ISBN 978-1-60618-993-1
1. Natural selection—Research—Great Britain—History—20th century.
2. Natural selection—Research—United States—History—20th century.
3. Biologists—Great Britain. 4. Biologists—United States. I. Title.
QH375.B78   2009
576.8'2—dc22   2009020654

FRONT COVER ILLUSTRATION: R. A. Fisher, from *Collected Papers
of R. A. Fisher*, edited by J. H. Bennett (University of Adelaide,
1971). Used by permission of the University of Adelaide.

CONTENTS

....

NO SCIENTIFIC THEORY IS WORTH ANYTHING unless it enables us to predict something which is actually going on. . . . There are still a number of people who do not believe in the theory of evolution. Scientists believe in it, not because it is an attractive theory, but because it enables them to make predictions which come true.

Haldane 1940, 7–8

THE FACT THAT NATURAL SELECTION offers a general explanation of adaptation is one of the chief reasons for the rapid acceptance of Darwinian theory among biologists. For adaptation is very widespread, and some of it is very remarkable. So abundant is it, and so marvelous are parts of it, that many naturalists have come to feel that adaptation is the outstanding feature of life requiring an explanation.

Shull et al. 1941, 359

1

....

Introduction

When did biologists accept Charles Darwin's theory of evolution? One might argue that his theory, in its original form, was never accepted by more than a handful of biologists. The theory that was accepted in the twentieth century was based on a different understanding of natural selection, of the variations on which it acts, and on the source of those variations. Nevertheless, by 1970, Darwin was given credit for proposing a theory that was *essentially* correct when the necessary corrections were made. In particular, biologists came to believe that natural selection acting on small mutations, combined with Mendelian principles of heredity, is the major cause of evolutionary adaptation. Additional factors, such as genetic drift, geographical isolation, polyploidy, and so forth, may be involved in some cases (especially in speciation) and at some levels, but they would accomplish little by themselves. As modified in the twentieth century, Darwin's theory, sometimes (misleadingly) called neo-Darwinism,[1] is generally believed by historians of biology to have been established in the 1940s. It was the core of an "evolutionary synthesis," which brought together previously separate disciplines such as genetics, zoology, botany, and paleontology into a unified science of biology. At the same time, it could be considered a constriction rather than a synthesis, because it excluded many possible causes of evolution (Provine 1988). There is now a growing scholarly literature on the history of this synthesis.[2]

Why did biologists accept the modern version of Darwin's theory? I

1. As noted by Dobzhansky (1942) and Bowler (2003), "neo-Darwinism" was the name used around 1900 to designate the theory of Weismann, which is only superficially similar to the theory of natural selection developed by Fisher, Haldane, and Wright (Gayon 1993). "Darwinism" itself is often misleadingly used in popular writings and in reception studies by historians to mean *not* Darwin's theory of natural selection but simply "evolution" (e.g., Glick 1974). I will therefore avoid both of these terms except in quotations.

2. Mayr and Provine (1980, 1998); Delsol (1991); Smocovitis (1996); Gayon (1998); Gerson (1998); Beurton (2002); Bowler (2003); Shanahan (2004); and other works cited next. The phrase "evolutionary synthesis" comes from the title of Julian Huxley's book (1942); see comments by Junker (2002) on the terminology.

....

discuss in this book the reception by evolutionists and other biologists of what I will call the *natural selection hypothesis* (NSH): *the hypothesis that natural selection, with an ample supply of variation in heritable characters, is not only the major process involved in evolution (with the help of geographical isolation or polyploidy in some cases), but also that Lamarckian effects, random genetic drift, and macromutations have essentially no evolutionary significance.* The NSH was accepted by a bare majority of evolutionary biologists for only a short time (in the 1950s and 1960s). In particular I ask, what were the reasons they gave for adopting that hypothesis and for rejecting the argument that drift plays a significant role in evolution?

Although I have generally used the concepts and terminology of evolutionary biologists in the time period under discussion, these concepts and terms may be confusing to a modern reader. Thus, John Endler argues that natural selection should be considered a *process* rather than a *cause*. It does not "act"—rather, it is the result of heritable differences and occurs when variation among individuals arises due to mutation or other causes.[3]

This study is part of a program to compare the reasons why scientists in different fields accept (or in some cases reject) proposed new theories. Is it true that "the reception of neo-Darwinism was colored by prevailing empiricist currents of thought" (Depew and Weber 1985b, 240)? If so, what specific empirical evidence was most persuasive? What weight did scientists give to the confirmation of predictions, supposedly the essential desideratum of the hypothetico-deductive scientific method, and according to Karl Popper, the basis of the criterion for demarcating science from pseudoscience?[4]

That question is especially appropriate in the case of evolutionary biology, since Popper's own critique of Darwinism provoked a long-running debate among philosophers and biologists about whether biological theories

---

3. Endler (1986, 4, 29; 1992). Natural selection is a *fact*, not a *cause* of anything (Wells, Huxley, and Wells 1931, 1:604). This may be set against the view of philosopher Elliot Sober (1984a, 27, 141), supported by Reisman and Forber (2005), that natural selection is a "force" (or "cause") that acts together with mutation, drift, and other forces. Dunn (1967, 49) is one of the few biologists who use that language; see also the 1944 comments by Stebbins in J. Cain (2004, 68). For further discussion of selection-type theories in biology and other sciences, see Darden and Cain (1989); Skipper (1999); Matthen and Ariew (2002); Nanay (2005); Skipper and Millstein (2005); Glymour (2006); Pigliucci and Kaplan (2006); and Plutynski (2007a).

I am grateful to an anonymous referee who suggested that I consult John Endler's excellent book, *Natural Selection in the Wild* (1986), even though as a primary source, it is well outside the time period covered by this book.

4. Popper (1934, 1959, 1962) and Brush (1995 [and references cited therein], 1996, 1999a, 2002). Here I use the word "prediction" in its nontechnical sense, corresponding to what physicists and some other scientists call "prediction in advance" and to what some philosophers call "novel prediction." See next note for further discussion.

should be expected to make testable predictions, and whether more generally, the methodological standards for biology should be different from those of the physical sciences. The creation of "philosophy of biology"—as distinct from traditional "philosophy of science," allegedly biased toward physics—resulted in part from this debate.

A remarkable aspect of the debate about predictiveness in biology is that its participants rarely mentioned the predictions, some of them confirmed, actually made by evolutionary theorists. Perhaps they assumed that if biological theories don't need to make testable predictions, it doesn't matter if they do so. Or one might argue that the basic Popperian premise—that real scientific theories are always judged by the results of their predictions, rather than, for example, by their ability to provide plausible and consistent explanations—is wrong.

Another problem with the emphasis on predictiveness is that the person who proposes a theory may not agree that a particular observation made by someone else would be a fair test of that theory. As Thomas Kuhn (1962) and others have pointed out, it is rare to find genuine "crucial experiments" in science, the kind of test whose outcome would force one side or the other to abandon its theory. In the controversy between Fisher and Wright about the role of random genetic drift in evolution, Wright often rejected the relevance of alleged tests of his theory (which was itself evolving), even when the results seemed to be favorable. It does happen, however, that a scientist will abandon a favorite theory because of the unexpected failure of his or her own experiment to confirm it.

In the other biological case I studied, Thomas Hunt Morgan's chromosome theory of heredity, geneticists considered that the most convincing evidence was Calvin Bridges' discovery of nondisjunction, yet they never mentioned that this was in fact a confirmed prediction (Brush 2002). Even Popper (1978) eventually retracted his statement—that Darwinian evolutionary theory is not scientific because it does not make testable predictions—but still did not acknowledge that such predictions had been made and tested. Conversely, Popper's paradigmatic example of prediction-testing in physics—the discovery of gravitational light bending, predicted by Albert Einstein's general theory of relativity—also contradicts his thesis, because physicists and astronomers gave at least equal weight to the successful calculation of another effect, the advance of the perihelion of Mercury, which had been observed for more than half a century and thus was clearly not a novel prediction.[5] In fact, I have not yet found any examples in twentieth-

5. Popper (1962, 117, 339–40; 1974, 36–37) and Brush (1989, 1999b). Physicists generally use the term "prediction" to include the deduction of facts already known and use "prediction

century physics in which novelty was generally thought to enhance the evidential value of a confirmed prediction, although it seems to have some value as a secondary criterion in chemistry (Brush 1996, 1999a).

Philosophers of science may wonder why I have focused on prediction to the exclusion of other philosophical issues relevant to natural selection.[6] I am not concerned with philosophical issues per se, but only to the extent that scientists themselves worry about them and perhaps take them into account in evaluating new ideas. This was certainly the case with predictiveness, which is a major part of the so-called scientific method as defined by many scientists (whether or not they follow it themselves). It is so widely believed that falsifiability is an essential property of a scientific theory that even the U. S. Supreme Court has cited it in justifying a landmark decision (Blackmun et al. 1993).

Another surprising thing about historical and philosophical studies of Darwin's theory is the huge quantity of literature on its reception in the nineteenth century, when it was *not* accepted (although evolution itself was widely accepted), compared with the much smaller number of studies dealing with it in the twentieth century, when it *was* accepted. Among reception studies by historians of science, Darwinism is one of the "big

---

in advance" when they want to specify that the facts were *not* known when the prediction was made; the latter is what philosophers of science call "novel prediction." Elie Zahar redefined "novel fact" to mean a fact that "did not belong to the problem-situation which governed the construction of the hypothesis" that predicted it (Zahar 1973, 103), but I will not need to use this definition. Some recent publications by philosophers of science have undermined the credibility of the "predictivism" thesis and have shown that explanation of a known fact (misleadingly called "accommodation") may be as good as, or even superior to, novel prediction. See Hitchcock and Sober (2004) and Scerri and Worrall (2001). Peter Lipton, a leading predictivist, conceded that the superiority of novel predictions is normative rather than descriptive; that is, he did not claim that scientists themselves give extra credit for novelty (Lipton 2005). See the book by Losee (2005) and its review by Allchin (2006) for further discussion of the relevance of "falsification" and "predictiveness" to the behavior of scientists.

Biologists, as will be seen below, do not have a single clear-cut position on prediction. John Patterson and W. S. Stone wrote: "As with all scientific theories, the validity of an evolutionary theory and the mechanisms postulated to explain it are tested in terms of their effectiveness in predicting *what has or will happen*" (Patterson and Stone 1952, 2; emphasis added). Claude A. Villee, noting that "there are some who claim that biology is not a science because it is not completely predictable," replies that even physics, "generally regarded as the most 'scientific' of the sciences, is far from completely predictable. . . . We cannot make predictions in the field of quantum mechanics, nor can we predict an earthquake" (Villee 1954, 4; 1957, 5). A physicist would point out that quantum mechanics does make very accurate predictions of atomic properties that can actually be observed, but it was originally accepted because it predicted facts that were already known but not explained; its "predictions in advance" were confirmed only after the theory was accepted.

6. Hull (1974); Sober (1984a, 1984b, 2006); and recent issues of *Studies in History and Philosophy of Biological and Biomedical Sciences* and of *Biology and Philosophy*.

three" theories—the others being Einstein's relativity and Freud's psycho-analysis.[7] Almost all of the Darwinian reception studies stop at or before 1900 (unless they cover the Scopes Trial and other aspects of the creation versus evolution debate), and they usually discuss evolution in general, with little attention to natural selection in particular. By contrast, I am not concerned here with the acceptance of the *fact* of evolution by biologists, which was almost universal in the twentieth century (see Numbers 2004), but with their debates about how evolution works.

Natural Selection in the 1930s

I start the story at the beginning of the 1930s, with the publication of the three canonical works of modern evolutionary theory by Ronald A. Fisher, J. B. S. Haldane, and Sewall Wright.[8] At that time, natural selection was generally recognized as a factor in evolution but was not considered suf-ficient by itself to account for the observed facts. It was sometimes argued that natural selection is only a *negative* factor: it can eliminate bad genes but does nothing to create good genes.[9]

George Parker, Director of the Zoölogical Laboratory at Harvard Uni-versity, asserted in 1925 that "most modern evolutionists . . . have one by one come to the conclusion that natural selection, which in Weismann's time was declared to be all-sufficient in evolution, may after all be of little real significance." Parker favored de Vries's theory of large mutations, on which natural selection could act, possibly with the help of a little Lamarckism.[10] But Lamarckism, though supported in England by E. W. MacBride (1929), was losing credibility. A weak version was periodically revived, then and

---

7. A search of the online version of the *Isis Cumulative Bibliography* (through the Research Libraries Information Network electronic database) using the words "reception," "response," and "reaction" suggests that about 20 percent of reception studies are devoted to Darwinism, 18 per-cent to relativity, and 15 percent to Freudian psychoanalysis; a more comprehensive search of the humanities and social science literature would probably increase the proportion of Freud and Darwin studies. In any case, these three have now become established as canonical cases for reception studies in the history of science (Glick and Henderson 2001).

8. For personal data see table 1 (p. 131). For a revisionist history emphasizing the earlier works of Fisher (1918) and Haldane (1924), see Sarkar (2004).

9. Morgan (1932, 130–1); G. E. Allen (1968, 1980); Punnett (1938, 221 [and other work dis-cussed by Kimler 1983b, 301–2]); Dalbiez (1927, 189); Lindsey (1931, 70, 113); Newman (1932, 395–6); Osborn (1932, 60); and McKerrow (1937).

"An analysis of textbooks and of the evolutionary literature shows that as late as 1930 se-lectionism was still a minority viewpoint in both Britain and the United States" (Mayr 1998, preface). Unfortunately, Mayr gave no details of this analysis. Lack (1957) cites a number of objections to natural selection.

10. Parker (1925, 113–44; see especially 119).

later, under names like the "Baldwin effect" or "canalization."[11] Saltationism (de Vries's large mutations) and several versions of orthogenesis were still advocated by some biologists.[12]

A good authoritative summary of the uncertain situation can be found in the Hooker Lecture, delivered to the Linnaean Society of London on March 14, 1929, by E. J. Allen, F.R.S., F.L.S., Director of the Marine Biological Association of the United Kingdom. Allen stated that there was plenty of evidence for environmental influence on the heredity of organisms, including Heslop Harrison's work on the production of melanic forms in moths and H. J. Muller's x-ray experiments on *Drosophila*. Yet Lamarckian theories of this influence had been discredited, and orthogenesis failed to provide the flexibility needed to adapt to a changing environment. This situation would seem to have provided an excellent opportunity for an explanation by natural selection, but that theory had so far failed to provide a plausible account.[13]

So in the 1930s, there was no widely accepted alternative to natural selection but instead a need for that process to be clarified and elaborated in order to serve as the major process in evolution. H. J. Muller's work on radiation-induced mutations and Dobzhansky's research on natural pop-

11. On canalization, see Waddington (1957, 1960); on "stabilizing selection," see Schmalhausen (1949); see discussion by Simpson (1953b), Richards (1987), and Lustig (2004).

12. On the pre-1930 arguments about whether large or small mutations are more important for evolution, see Olby (1981); Gould (2002, 425–50); Bowler (1978; 2003, 268–72, 327–30); and Provine (1971, 13, 56–57, 64–69, 91–92). According to Ernst Mayr, in the 1920s the leading Mendelians (Bateson, DeVries, Johannsen, Goldschmidt, and Morgan) were "saltationists who had little use for natural selection" (Mayr 1992b, 179).

On orthogenesis, see Bowler (1979; 1983, 144, 162, 178–81); Holmes (1948a); Jepsen (1949); and Richardson and Kane (1988).

13. "Can the germ-plasm itself be directly or indirectly acted upon by physical or chemical changes in the environment in such a way that variations, or, to be more precise, mutations, are produced in the offspring—variations which are capable of being transmitted to future generations[?] Well-substantiated evidence is now, I think, forthcoming that the germ-plasm can be so acted upon . . ." (Allen 1930, 133). Allen asserted that Lamarckian theories, despite the support of MacBride, now seemed implausible, especially in the light of Graham Kerr's argument: "Is there any possible escape from the conclusion that, were it actually the case that impressed characters are transmitted, we should have before us a mass of evidence so overwhelming as to leave not the slightest doubt in any one's mind as to the occurrence of such transmission" (Kerr 1926: 62 [as quoted by Allen 1930, 133]). "Natural Selection, as conceived by Darwin and Wallace, still seems to most competent biologists a potent factor, but the number who regard it as all-sufficient would seem to be diminishing. We must even face fearlessly the possibility that the whole conception must be abandoned as a main cause of organic evolution" (Allen 1930, 137). Yet Graham Kerr, in the book just cited, made a rather strong case for natural selection and refuted several criticisms of it.

ulations helped to broaden the definition of natural selection to include variation as a basic component rather a separate factor.[14]

In their development of the natural selection hypothesis, Fisher, Haldane, and Wright used different mathematical approaches based on similar assumptions to arrive at results that were, with some minor exceptions, in good agreement with each other. Fisher and Wright disagreed on the relevance of random genetic drift to the process of evolution as it actually occurred in nature, and this disagreement was perceived as a controversy by Fisher and others. However, by 1960 Wright had retreated from his earlier pro-drift position and asserted that he had not earlier and did not now deny the primacy of natural selection. (He continued to believe in his "shifting balance" theory, involving random drift in subdivided large populations, up to his death.)

In this book, I focus on the competition between natural selection and drift as perceived not only by Fisher and Wright but also by others in that period; I do not discuss in detail the shifting balance theory or the revival of drift after 1965. This focus may leave the reader with an unfairly negative impression of Wright's contributions, an impression that should be remedied by looking at the publications of Provine and Skipper. Moreover, the NSH considered here was rather simplistic compared with later discussions of "levels of selection"; I avoid many of the more sophisticated arguments made about natural selection by biologists and philosophers in recent decades, because they did not seem to play an important role in the acceptance of natural selection itself before 1970.

By the mid-1960s the other major alternatives to selectionism had been discredited, so one may say that the strong selectionist view had become established knowledge among the leaders of evolutionary theory. (This is what Stephen Jay Gould [1983] called the "hardening" of the synthesis.) So we can then ask the question: when, and to what extent, did this knowledge (the primacy of natural selection) infiltrate the writing and teaching of other biologists?

Unlike some of the other theories I have studied, which remained orthodox doctrine for at least a couple of decades, selectionism was challenged from several directions beginning in the late 1960s (Dietrich 1994). I have taken 1970 as my endpoint in order to keep this monograph from becoming impossibly long. I do not discuss later research on evolution, ex-

---

14. Muller (1927); Patterson and Muller (1930); and Dobzhansky (1981). According to Roman (1988), one should also give some credit to Lewis J. Stadler (1928) for his almost simultaneous discovery.

cept for historical and philosophical writings relevant to the period before 1970.

This study is limited in space as well as in time: it covers only the Anglo-American biological community. Of course this community includes prominent evolutionists such as Theodosius Dobzhansky and Ernst Mayr, who emigrated from other countries in the 1930s. But it does not include Russia, Germany, and France; one may obtain a general understanding of what was going on there from some excellent recent historical studies.[15] Another limitation, not so easily remedied: I have mostly used published works, rather than attempting to examine all the correspondence of the hundreds of biologists whose views on natural selection might be relevant. Also, I have omitted some biologists who made important contributions to the evolutionary synthesis but were not directly involved in research on natural selection; Edgar Anderson is an example. I have not examined the laboratory notebooks of biologists who, as an anonymous referee of this book suggests, may have been testing predictions without saying so in their published reports, or conversely, may have been "reconstruct[ing] investigations in an overly rational way" in their publications so as to make them look like tests of predictions, "when in fact they had other designs." I can only hope that other scholars, more acquainted with individual scientists and their archives, will fill in this gap.

15. For France, see Buican (1983, 1984, 1989) and Grimoult (2000, 2001). For Germany, see Brömer et al. (2000); Harwood (1985); Hossfeld (2000); Junker (2000, 2002, 2004); Junker and Engels (1999); Junker and Hossfeld (2000, 2002); Laubichler (2006); Mayr (1999); and Reif et al. (2000). For Russia, see Adams (1970) and Kolchinsky (1999, 2000a, 2000b). Krimbas (1995) provides a comprehensive survey of the direct and indirect personal influences of Dobzhansky on biologists in several countries; see also the other articles in the same book (Levine 1995). Jonathan Harwood writes that all of these works refute "claims that the evolutionary synthesis was largely an Anglo-American achievement" (Harwood 2002, 369–70).

2

....

Fisher: A New Language
for Evolutionary Research

The early reception—or perhaps nonreception—of the modern theory of natural selection was strongly conditioned by the presentation of the theory in a language incomprehensible to most Anglo-American biologists. I am not referring to the fact that some important features of the evolutionary synthesis were originally published in Russian by Sergei S. Chetverikov (1926) and his colleagues. Rather, I mean that Fisher especially, and to a lesser degree Haldane and Wright, reasoned and wrote in the language of nineteenth-century theoretical physics. That language cannot easily be translated into the language of early twentieth-century biology, because as in understanding any language, one cannot simply look up individual words in a dictionary; one must learn something about the background knowledge and style of thought of those who use the language.[16]

16. As noted by J. F. Crow, among the founders of the evolutionary synthesis, "Fisher was far and away the most skilled and inventive mathematically. He solved very difficult problems, and almost always with elegance and grace." His *Genetical Theory* "is the most important book on evolution since Darwin" (Crow 1992c, 99); see also Crow (1990b) and Edwards (2000). See also the letter from Fisher to M. Kimura, May 3, 1956, in Bennett (1983, 229–30). But "his papers could not be understood by most mathematicians, with the result that they were often not trusted. . . . Fisher hardly ever made it clear what his assumptions were, when and how he was approximating, and how to get from one equation to the next" (Crow and Dove 2000, 142, 144). A more dangerous aspect of Fisher's work is what Haldane considered a desirable result: "Discussions of Darwinism by persons wholly ignorant of mathematics will come to be classed with circle-squaring and earth-flattening" (Haldane 1931, 475). See also Ford (2005).

As hypothetical history, one might ask how the development of Darwin's theory and of post-Darwinian biology would have been different if he had studied the works of Adolphe Quetelet, including Quetelet's critique of Malthus's law of population growth, and had adopted a statistical approach to natural selection (Ariew 2007). Instead, most biologists resisted mathematical methods until after World War II. This is (to me) illustrated by the great significance they gave to the "Hardy-Weinberg law," which, as James Crow points out, "is so self-evident that it hardly needed to be 'discovered'" (1988, 473).

....

A successful theoretical physicist often chooses a simple model, deliberately ignoring the rich complexity of the natural world so dear to the naturalist, in favor of a set of postulates whose mathematical consequences can be reliably computed. It may be helpful to note that a theoretical *model* does not make empirical claims, but a *scientist* may make the empirical claim that a particular phenomenon is more or less well described by the model. Modern evolutionary theory based on the models developed by Fisher, Haldane, and Wright can be interpreted as an example of the "semantic view of theories," now popular in the philosophy of science, rather than the (no-longer) "received" view of theories, according to John Beatty (1981).

In Fisher's case it is essential to note one biographical fact: he studied theoretical physics, including the kinetic theory of gases, with James Jeans.[17] Hence the following sentence, which appears halfway through the first chapter of *The Genetical Theory of Natural Selection*:

> The particulate theory of inheritance resembles the kinetic theory of gases with its perfectly elastic collisions, whereas the blending theory resembles a theory of gases with inelastic collisions, and in which some outside agency is required to be continually at work to keep the particles astir. (Fisher 1958, 11)

In the particulate theory, "elastic" interactions tend to maintain a statistical distribution (of speeds, or genes). The combination of many random causes may produce a single clear-cut effect.[18] Fisher's statistical approach implied that fluctuations from an average value become relatively insignificant for a large population, just as the pressure of a gas at a fixed temperature in a fixed volume remains constant unless the density is extremely low. So when A. L. Hagedoorn and A. C. Hagedoorn proposed that random mating of organisms has a greater effect than natural selection, Fisher immediately objected that this "Hagedoorn effect" (a version of Sewall Wright's random

17. Box (1978, 33) writes that Fisher completed the Cambridge Tripos examination in June 1912 as "a Wrangler with distinction in the optical paper in Schedule B. . . . Because he had been awarded a studentship in physics, he was able to return to Cambridge in October for a year of graduate study of statistical mechanics and quantum theory under James Jeans and the theory of errors under F. J. M. Stratton." Jeans authored a standard text (*The Dynamical Theory of Gases*) and popular books on science, contributed to the theory of black-body radiation, and proposed a theory of the origin of the solar system. Fisher also compared Darwinian natural selection with the kinetic theory of gases in a 1927 lecture at Leeds, a summary of which is found in the Fisher papers, Leeds file (I thank Nancy Hall for this information). For further discussion of the influence of statistical physics, see Hodge (1992); Depew and Weber (1995, chapter 10); and Morrison (1997, 2000, 2004). His published papers are reprinted in Fisher (1971–1974).

18. Fisher and Stock (1915, 60) and Fisher (1922, 321).

genetic drift, discussed next) must be insignificant except in small populations (Fisher 1921, 1922).

Fisher (1930a) had already pointed out that in Darwin's theory based on blending inheritance,

> bisexual reproduction will tend rapidly to produce uniformity [and therefore] if variability persists, causes of new variation must be continually at work; [and hence] the causes of the great variability of domesticated species, of all kinds and in all countries, must be sought for in . . . changed conditions and increase of food.[19]

So, the variability must be maintained by frequent mutations and/or environmental changes in a theory based on blending inheritance.[20]

Preserving or increasing the variability was essential to evolution because of what Fisher called the "fundamental theorem of natural selection." Fisher's original proof of the theorem was muddled, and the theorem itself is literally incorrect as printed; evolutionary biologists simply accepted it as qualitatively correct, with no serious technical discussion until many years later. I will therefore simply state the original without commenting on its derivation: "The rate of increase in fitness of any organism at any time is equal to its genetic variance in fitness at that time."[21]

According to George R. Price (1972), the theorem is correct if one defines "change in fitness" as change in the *fraction of total change* due to natural selection, not including the effects of environment, and if one corrects a large number of errors and ambiguities. There have been many other attempts to explain what the theorem means and how valid it is,[22] but for my purposes, the qualitative statement that the speed of evolution is an increasing func-

19. The words in brackets were inserted in the 1958 edition; see Fisher (1999). Fisher evidently felt the need to emphasize that these statements, originally presented as a list of separate postulates, were logically connected.

20. He did not mention Maxwell's velocity distribution, much less discuss its variance-preserving property, but probably took it for granted that anyone familiar with the kinetic theory of gases (e.g., anyone who studied for the Cambridge University Mathematical Tripos examination) would immediately see the point of the analogy between gene frequencies and molecular velocities. See Garber et al. (1986, 9–11, 281–2, 288–95, 320–1).

21. Fisher (1999, 35). The theorem as stated is dimensionally incorrect, since it has a time variable on the left but not on the right; presumably, one is supposed to assume that time is measured in number of generations. For further discussion of the theorem, see Price (1972); Olby (1981); Endler (1986, 28–29); Leigh (1986); Ruse (1996, 316); Crow (2002); and Plutynski (2006). Ayala (1965, 1968a) provided a qualitative experimental verification.

On the ambiguity of the concept of fitness, see Depew and Weber (1995); Ariew and Lewontin (2004); Sober (2006 [reprinting articles by S. K. Mills and J. H. Beatty and by Sober, 3–39]).

22. Lessard (1997); Plutynski (2006); and works cited by these authors.

. . . .

tion of the amount of genetic variance is sufficient; it is not necessary to undertake a technical discussion of the theorem here, since the reception of natural selection before 1970 did not much depend on the theorem.

Fisher introduced another important feature of the modern theory of natural selection when he called attention to the misleading use of the phrase "struggle for existence" in the evolutionary literature. This struggle was usually blamed on the "excessive production of offspring, supposedly to be observed throughout organic nature" (Fisher, 1930a, 46). This manner of speaking, which places the emphasis on overpopulation as a cause of competition between individuals, diverts attention away from the importance of understanding the quantitative aspects of reproduction:

> There is something like a relic of creationist philosophy in arguing from the observation, let us say, that a cod spawns a million eggs, that *therefore* its offspring are subject to [n]atural [s]election; and it has the disadvantage of excluding fecundity from the class of characteristics of which we may attempt to appreciate the aptitude. It would be instructive to know not only by what physiological mechanism a just apportionment is made between the nutriment devoted to the gonads and that devoted to the rest of the parental organism, but also what circumstances in the life-history and environment would render profitable the diversion of a greater or lesser share of the available resources toward reproduction.[23]

Fisher's view of the operation of selection, endorsed by Dobzhansky (1942, 397), foreshadowed the general shift in the conception of "evolutionary fitness" away from survival of the individual organism in a brutal battle with other organisms to a definition of fitness in terms of differential reproductive success (similar remarks may be found in Thomson 1920). According to Peter J. Bowler (2004), T. H. Morgan's opposition to Darwinism "was dictated by his moral objections to the picture of a world dominated by struggle" (Bowler 1978, 55).

The shift away from a "struggle for survival" to "struggle to reproduce" was to prove useful in combating the harsh image of natural selection, blamed by some critics for the excesses of the Nazis as well as for "social Darwinism." Tennyson's famous line "Nature, red in tooth and claw" (1850, section LVI, fourth quatrain) was frequently quoted. Less memorable but more appropriate was T. H. Huxley's statement in his 1893 Romanes Lecture: Survival of the "fittest" does not mean survival of the "best" in the ethical sense. On the contrary, "The struggle for existence tends to elimi-

---

23. Fisher (1930a, 47); cf. Fisher (1930b).

nate those less fitted to adapt themselves to the circumstances of their existence. The strongest, the most self-assertive, tend to tread down the weaker" (T. H. Huxley 1947, 81–82). But in the 1960s, some American textbook writers invoked the counterculture slogan "Make love, not war" by calling natural selection a "peaceful force" that promoted sex rather than a destructive one that promoted death.[24]

Moreover, sexual rather than asexual reproduction is quantitatively important to evolution, according to Fisher's theory. He pointed out that while natural selection would allow an asexual organism to evolve, an otherwise similar sexual organism would evolve much faster, depending on "the number of different loci in the sexual species, the genes in which are freely interchangeable.... Even a sexual organism with only two genes would

---

24. Moody (1962, 355); Rhodes (1962, 277); Savage (1963, 50); C. Wallace (1969, 125); Weisz (1965, 430; 1966, 352–4; 1968, 196). According to Gerson (1998, 356), this shift began after World War I. According to Ernst Mayr's unpublished "Travel Notes, 1954": "In Germany—now a clerical state—the anti-evol[utionary] movement is particularly strong.... Just like McCarthy synonymizes liberalism and communism, thus after the war evolution was synonymized with the most typological selectionism, and biology with Nazi racism" (quoted by Junker 2002, 225). For a contrary view, see Huxley (1963, introduction) and his letters to Simpson published in Dronamraju (1993, 240–3).

C. H. Waddington, commenting on the statement of T. H. Huxley quoted in the text, wrote that "he was writing under the spell of that extraordinary impulsion, so incomprehensible to us today, which forced the Victorians to transmute the simple mathematics of the major contribution of theoretical biology into a battle-ground for their sadism" (Waddington 1942, 17). Waddington rejected the conclusion that evolution leads to "an increase in bloodiness, fierceness and self-assertion." On the contrary, he asserted, "We must accept the direction of evolution as good simply because it *is* good according to any realist definition of that concept" (Waddington 1942, 18). Karl Popper, "feeling somewhat intimidated by the tendency of evolutionists to suspect anyone of obscurantism who does not share their emotional attitude toward evolution"—exemplified by Waddington's assertion—conceded, "I see in modern Darwinism the most successful explanation of the relevant facts" (Popper 1960, 106–7). This concession did not prevent Popper from denying the scientific status of Darwinism (see chapter 21).

Fisher (1947) ascribed the late nineteenth- and early twentieth-century opposition to Darwinism to "psychological sources": it was felt to be "a harsh and ruthless sort of theory" and an example of materialistic determinism. But now, he wrote, the physicists have liberated us from determinism and we can recognize that natural selection is a creative process whose causes "lie in the day to day incidents in the innumerable lives of innumerable plants, animals and men"—not "set in motion on 'the first morning of creation' or determined by heredity or environment" (Fisher 1947, 619, 620); see also Fisher (1950).

Fisher's concern with reproduction is of course directly related to his interest in eugenics, which according to B. Norton (1978, 1983) was a motivation for his research on natural selection; it takes up about a third of the text of his *Genetical Theory*. Mayr remarks: "Curiously, factors that contribute to reproductive success were largely neglected by evolutionists until around 1970. At that time naturalists rediscovered Darwin's important finding (1871) that females may play a decisive role in the choice of their mates" (Mayr 2004, 139). He attributed the earlier neglect to "adopting the gene rather than the individual as the target of selection," which "eliminated the difference between natural and sexual selection" (Mayr 1992a, 17).

. . . .

apparently possess a manifest advantage over its asexual competitor ... from an approximate doubling of the rate with which it could respond to [n]atural [s]election."[25]

A result that illustrates more sharply Fisher's mathematical sophistication is that large mutations are generally nonadaptive or lethal, so only small mutations can contribute to evolution. Fisher's reasoning was geometric: Let point A represent the position of the organism in an n-dimensional space, where each dimension represents the value of some attribute of the organism. (This is like the "phase space" familiar to students of the kinetic theory of gases.) Let point O, at distance d from A, represent the set of values of those attributes corresponding to the best possible adaptation to the environment. A mutation will be favorable only if it moves the organism closer to O. Suppose mutations are random (not correlated with adaptiveness). The probability that a mutation is favorable is simply proportional to the amount of space that is closer to O than A is.

For those who think in three-dimensional space, it seems reasonable that for a *very small* mutation—moving the organism a distance r away from A—the chance of ending up closer to O is nearly 50 percent, since the surface of the sphere centered at O is nearly a plane. The organism is slightly more likely to move farther away from the sphere. On the other hand, if r is *large* (compared to d), it may go through the sphere and land outside it—an unfavorable mutation. If r is *greater* than $2d$ (the diameter of the sphere), it cannot land inside the sphere; the probability of a favorable mutation is zero.

For someone like Fisher who thinks in n-dimensional space, it is obvious that the same argument applies with even greater force: the greater the number of dimensions (attributes of the organism), the smaller the probability that a large mutation will be favorable.[26] Of course it was precisely the effort to think outside the three-dimensional box of everyday experience that was one of the driving forces of early twentieth-century physics and art (Miller 2001).

As Fisher pointed out, the idea that most mutations are harmful is not

25. Fisher (1930a, 121–3). According to J. F. Crow, "Muller shares with R. A. Fisher (1930) credit for the first convincing statement of the value of Mendelian segregation and recombination," as illustrated by this explanation of sexual reproduction, even though "the Muller-Fisher idea is no longer regarded as the major reason for the evolution and maintenance of sexual reproduction, at least in multicellular eukaryotes, although it may have been at an early time in the history of life" (1992c, 85, 88 [citing Muller 1932]).

26. Fisher (1999, 38–39). It has been suggested that Fisher developed his facility for visualizing n-dimensional space because of his poor eyesight (Box 1978, 15). Crow (1990c) writes that "Fisher had an insight into multidimensional geometry that was little short of occult" (Crow and Dove 2000, 142).

original but was "regarded as obvious by the naturalists" who believed that organisms are generally very well adapted to their environments. It was the popularity of the de Vries large-mutation theory that made it necessary to reinforce that view.[27]

If natural selection acts primarily on very small mutations, it might seem that it must be a very slow process. But Fisher showed that a rare gene with only a small selective advantage—a small mutation that produces no observable effect in the first generation—can spread rather quickly through a population if the same mutation is repeated (randomly but at a finite rate) in every generation. For example, if the selective advantage is 1 percent (and the mutation rate is one in one thousand million [per generation]), then "in a species in which 1,000,000,000 come in each generation to maturity, a mutation rate of 1 in a thousand million will produce one mutant in every generation [on average], and thus establish the superiority of the new type in less than 250 generations, and quite probably in less than 10, from the first occurrence of the mutation, whereas, if the new mutation started with the more familiar mutation rate of 1 in 1,000,000 the whole business would be settled, with a considerable margin to spare, in the first generation." Natural selection will be much less effective if the population is very small.[28]

Fisher's theoretical conclusion that small mutations—those whose effects are not easily observable in one generation—are more important in evolution than large ones was confirmed by later experiments. An example was the study by Kenneth Mather (Fisher's first genetical assistant when he became Galton Professor at University College London) and L. G. Wigan. They found that the several small mutations that combine to determine a polygenic character, such as the number of chaetae (hair or bristles) on the abdomen of a *Drosophila* fly, produce an effect that is initially masked by fluctuations. Under the influence of selection, they may produce a continuous change or a sudden jump that mimics a macromutation (Mather and Wigan 1942; see also Mather 1943).

The quantitative effectiveness of natural selection was first revealed in calculations by H. T. J. Norton, published by R. C. Punnett in 1915 (Norton and Punnett 1915). In later years, many evolutionists used a phrase without citing a specific source but attributed to Fisher: "Natural selection is a mechanism for generating a high degree of improbability." J. S. Huxley reported that he "asked Professor Fisher where he had published this, to

27. Fisher (1999, 41). Turner (1985) discusses how Fisher handled the phenomenon of mimicry, formerly considered strong evidence for large mutations; see also Leigh (1987).

28. Fisher (1999, 78). At that time, the word "billion" was ambiguous; it meant 10^9 in the United States, but it meant 10^{12} in the United Kingdom, which subsequently decided to adopt the American definition.

which he replied that he had merely stated it verbally at a meeting. However, it sums up the situation so pithily that I feel it should be recorded in print" (J. S. Huxley 1951, 595). Later, Fisher printed his own version of the statement: natural selection is the "process by which contingencies *a priori* improbable are given, in the process of time, an increasing probability, until it is their nonoccurrence rather than their occurrence which becomes highly improbable."[29]

The above results show that natural selection may be an effective cause of evolution but does not appear to lend itself to empirical tests in a short period of time unless the selective advantage or the mutation rate is large. Fisher also discussed situations in which stable gene ratios are maintained by a balance of selective forces: "One gene has a selective advantage only until a certain gene-ratio is established, while for higher ratios it is at a selective disadvantage. In such cases the gene ratio will be stable at the limiting value, for the selection in action will tend to restore it to this value whenever it happens to be disturbed from it in either direction." Of course the conditions of stability must themselves be transient during the course of evolutionary change, but the temporary stability provides a good opportunity to study in detail how natural selection works. Theory indicates that "a single factor may be in stable equilibrium under selection if the heterozygote has a selective advantage over both homozygotes."[30] This situation, later known as a "balanced polymorphism," was investigated extensively by Fisher's Oxford colleague E. B. Ford and his students and provided important evidence for the validity of the theory.

While Fisher's theory is sometimes described as an application of Mendelian genetics to Darwin's theory of natural selection, he also proposed a significant change in Mendel's theory. The attribute of genes called dominant or recessive was, he argued, not fixed, but was itself a result of evolution (Fisher 1930a, chapter 3). Although Fisher himself considered this hypothesis to be of great importance, it seems to have had no influence on the reception of his general theory, with a few exceptions (e. g., J. Huxley 1947, 167). An excellent account of the hypothesis and its empirical test is given in chapter 8 of the biography by his daughter Joan Fisher Box, who concludes: "Among geneticists today the concept of the evolution of dominance is almost unknown. Nevertheless, it has played a significant part in transforming scientific opinion in respect of evolution by natural selection.

29. Fisher (1954, 91). I thank Will Provine for this reference. Huxley himself stated that "the enormous degree of that improbability was, I think, first clearly demonstrated in a general way by Muller in 1929," but he did not cite Muller's publication.

30. Fisher (1930a, 99–100); see Fisher (1922) for an earlier statement of this principle.

It has opened the whole field of ecological genetics by which the efficacy of natural selection has been fully demonstrated" (Box 1978, 233). Later research indicated that Fisher's theory of dominance is not supported by empirical data (Skipper 2000, 105–15).

In developing a new field, it is important to be able to publish one's results quickly, even when the editors and referees of established journals are indifferent or hostile. In 1947, together with the British cytologist Cyril Darlington, Fisher cofounded *Heredity: An International Journal of Genetics*. Although Darlington did most of the editorial work, Fisher "was content that Darlington was providing in *Heredity* a ready platform for the selectionist crusade of his protégé E. B. Ford and Ford's students Bernard Kettlewell, Arthur Cain, and Philip Sheppard." (Harman 2004, 209–10).

Readers seeking a "social construction" interpretation of Fisher's theory should consult the papers by Bernard Norton (1978, 1983), Robert Olby (1981), and James Moore (2007) for discussion of how Fisher's theory is related to his eugenic and religious views. They should also inspect Fisher's own text to check the accuracy of Norton's assertion (1983, 27) that Fisher was not interested in speciation because his motive was to improve one species (our own) rather than change it to another. In fact Fisher did discuss speciation, suggesting an analogy with the fission theory of the moon's origin proposed by Charles Darwin's son George Howard Darwin.[31]

In summary, Fisher was the most important of the three founders, not because he came first alphabetically or chronologically (his first major contribution was published in 1918), but because he established a simple proposition, subject to certain assumptions: natural selection acting on small mutations can produce a significant change in the genetic composition of a population much faster than was previously believed. Thus, a theory based on natural selection *may* be able to explain evolution and is therefore worth pursuing. Biologists criticized Fisher's model (as defined by his assumptions) for being unrealistic—"bean bag genetics."[32] But from the viewpoint of theoretical physics, that is not a criticism at all. It is often by starting with an idealized model whose properties can be accurately calculated that we can start to understand how nature works.[33] Of course it takes a while for

---

31. Fisher (1930a, chapter 6); Olby (1981, 256–7); and Brush (1996a, chapter 4.2).

32. The phrase apparently originated with Mayr (1959), who admitted that it did not apply to Wright, and that, as used by Fisher and Haldane, it was "a necessary step in the development of our thinking . . ." (Mayr 1992b, 21).

33. Lewontin (2000) and Plutynski (2001, 2004, 2006). Full disclosure on this point: in my former career as a theoretical physicist, I followed just this approach, working only with idealized models, whereas Fisher actually did some experiments to test his theories (e.g., Brush et al. 1966).

other scientists to be persuaded that this approach can work in biology, and so it was very fortunate that Fisher attracted followers like E. B. Ford who could apply and propagate his ideas.

The previous paragraphs may give the impression that Fisher was an armchair theorist who left it to other scientists to do the empirical work needed to test his theories. But he did take an active role in empirical research, by himself and in collaboration with E. B. Ford and others, as may be seen from the list of his papers on genetics, evolution, and eugenics at the beginning of his *Collected Papers*.[34] An early example is the confirmation of a prediction, suggested by Charles Darwin and quantitatively derived from Fisher's theory of natural selection, that the variability of a species should be proportional to the size of its population (Fisher and Ford 1926, 1928). According to Ford (1980, 338), Fisher's 1927 paper on mimicry was "the true start of the modern evolutionary synthesis," because it "pushed people into the field" to study natural populations; it "drew attention to the selective adjustment of the effects of genes."

Fisher's fame as a leader in the development of statistical methods in research undoubtedly helped to call attention to his development of the NSH; his method of double-blind random trials is often called the gold standard for testing new substances and procedures in medicine and agriculture.[35] For some historians and philosophers of science, his critical scrutiny of Mendel's data showed that one can never assume that empirical results are independent of the theoretical bias of the observer; Fisher's statistical analysis suggested that Mendel's published results are too good to be true.[36]

34. Fisher (1971–1974, 1:14–17).

35. N. L. S. Hall (2007) and Box (1978).

36. Fisher (1936a); for discussion, see Pilpel (2007), Franklin et al. (2008), and works cited therein.

3

....

Wright: Random Genetic Drift,
a Concept Out of Control

Sewall Wright, in his comprehensive 1931 article, showed that many of
his results were consistent with those of Fisher but stressed the importance of population size in evolution.[37] His earlier work on inbreeding, as
an employee of the U.S. Department of Agriculture, indicated that random fluctuations are important in small populations. They will generally
tend to decrease the *heterozygosity* (presence of different genes; e.g., A and
a at the same locus on corresponding chromosomes of an organism) at a
rate inversely proportional to the population size N. This means that traits
determined by the homozygous genes (e.g., A and A) will tend to become
fixed (Wright 1931, 158). In such a small population there is

> little variation, little effect of selection and thus a static condition modi-
> fied occasionally by chance fixation of new mutations leading inevitably to
> degeneration and extinction.

On the other hand, in a very large interbreeding population, there is

> great variability but such a close approach to complete equilibrium of all
> gene frequencies that there is no evolution under static conditions.

The most favorable situation for evolution is

> a large population, divided and subdivided into partially isolated local
> races of small size.... Complete isolation in this case ... originates new
> species differing for the most part in nonadaptive respects, but is capable

37. Wright (1931), reprinted with other papers in Wright (1986). See the comprehensive
biography by Provine (1986) and articles by Crow (1994), Hill (1990), and Park (1991).

....

of initiating an adaptive radiation as well as parallel orthogenetic lines, in accordance with the conditions. (Wright 1931, 158)

At the Sixth International Congress of Genetics (in Ithaca, New York, 1932), Wright presented a shorter, more qualitative version of his theory. This, according to his biographer William Provine, "was probably the most influential paper he ever published."[38] He introduced diagrams to show various hypothetical "fields of gene combinations." The species will generally occupy a region around a "peak" representing the optimum combination of gene frequencies in a particular environment. Isolated small local populations will evolve by random genetic drift:

> The chances are good that one at least will come under the influence of another peak. If a higher peak, this race will expand in numbers and by crossbreeding with the others will pull the whole species toward the new position. The average adaptiveness of the species thus advances under intergroup selection, an enormously more effective process than intragroup selection.[39]

Wright's general theory was called the "shifting balance" process of evolution. It is another example of introducing the language of nineteenth-century theoretical physics into twentieth-century biology: it invokes the "fields" used qualitatively by Michael Faraday to describe electrical and magnetic phenomena but without the mathematical sophistication of James Clerk Maxwell or R. A. Fisher.

Robert A. Skipper Jr. has given a comprehensive philosophical-historical analysis of Wright's theory and summarizes it as follows:[40]

Phase I. Random genetic drift causes subpopulations semi-isolated within the global population to lose fitness.

Phase II. Selection on complex genetic interaction systems raises the fitness of those subpopulations.

Phase III. Interdemic selection then raises the fitness of the large or global population.

38. Provine in Wright (1986, 97). Ruse (1996b) argues that Wright's use of diagrams, though not logically necessary, helped to persuade other scientists to adopt his theory. See also Skipper (2004).

39. Wright (1932); reprinted in Brousseau (1967, 68–78 [quotation on 75]).

40. Skipper (2000; 2004, 1177).

James Crow comments:

It is important to realize that Wright thought of his process as a way of going from one adapted state to another when intermediates are unfit. In his metaphor, random drift could occasionally cause the population to drift across a valley to a point where it came under the domain of attraction of a higher peak. The contrast between Fisher and Wright is that Wright regarded the inability of mass selection to carry a population from one stable state to another whereas Fisher believed that hardly ever would a population be in a position such that *no* change in allele frequencies could cause an increase in fitness. Fisher thought that Wright's peaks and valleys were more like waves and troughs in an ocean. . . . When Wright says intergroup selection he really had in mind directional migration between subgroups.[41]

This mechanism for evolution following isolation, which Wright called "an essentially nonadaptive one" although it involved selection as well as genetic drift, became known as the "Sewall Wright effect."[42] According to Provine (1983), Wright originally favored genetic drift because it was used by several evolutionists to explain nonadaptive differences between similar species.

Despite Wright's frequent statements that he did not think genetic drift by itself would produce evolution in the absence of selective forces, his ef-

41. Crow, commenting on an earlier version of this book, reviewed for the American Philosophical Society, June 2005; see also Crow et al. (1990).

42. Wright (1986, 168–9) and Provine (1983, 1985, 1986). Provine credits the Hagedoorns with pointing out the importance of random fixation in evolution in 1921 (A. L. Hagedoorn and A. C. Hagedoorn 1921; Wright 1986, 108). Apparently unknown to Wright, N. P. Dubinin (1931) recognized a similar effect (see also Dubinin and Romashov 1932), and it was subsequently occasionally called the Wright-Dubinin effect. See also Wright (1931); Mayr (1963, 204); and Dobzhansky (1970, 231). Even earlier, H. J. Muller (1918) introduced "drift" under a longer name. See his paper in Huxley (1940, 185–268), partly reprinted in Muller (1962); and see Lewontin (1974, 87–90).

Here's an example of an early textbook explanation: "The accidental meeting and pairing of individuals will usually result in some deviation from the expected result. A gene that ought theoretically to occur in 25 per cent of the individuals may easily happen to be in 28 per cent or only 22 per cent solely by chance. Should the deviation from expectation in the next generation happen to be in the same direction, the difference is accentuated. Different parts of a range may thus come to be inhabited by groups of individuals which, while still belonging to the same species, nevertheless have their genes in different proportions. . . . Possibly even a divergence great enough to mark two separate species may take place in this purely random manner. Beyond this degree of differentiation probably other factors enter. The most important of such factors is believed to be natural selection." (Shull et al. 1941, 355).

fect and his term "drift" were most often invoked to explain (supposedly) nonadaptive characters (Lack 1940, 324). Wright himself may have encouraged this incorrect usage by statements such as that the distribution of the three allelomorphs for human blood groups is "apparently nonadaptive" (Wright 1931, 127) and that

> in the human species, the blood group alleles are neutral as far as is known. The frequencies vary widely from region to region and in such a way as to indicate that the historical factor (i.e., partial isolation) is the determining factor. The frequency distribution indicates a considerable amount of random differentiation even among the largest populations. (Wright 1940, 179)

The refutation of these statements was later used, perhaps unfairly, as a refutation of Wright's entire theory of evolution (Provine 1986, 456).

A further confusion arose from the statistical nature of natural selection itself. If one defines "fitness" as reproductive success, then it is difficult to distinguish between drift as a sampling error and natural selection.[43]

Wright's shifting balance theory was fairly popular in the United States before 1950, partly because it was adopted by Dobzhansky and partly because his inbreeding formulas were found to be useful in improving agricultural production. For example, poultry breeders were able to increase the number of eggs laid by hens, an achievement that makes Wright indirectly responsible for the fact that eggs are relatively cheap in the United States today.[44] Although Wright's contribution to statistical method was small compared to Fisher's, his "path analysis" technique became popular in the social sciences in the 1970s (Crow 1990a).

The career of genetic drift reached a high point in 1954 when Wright published a series of three papers reporting experiments done by Warwick E. Kerr, a visitor from Brazil supported by a Rockefeller Foundation fellowship. Very small populations of *Drosophila melanogaster* (four males and four females in each) were isolated and followed through ten or more generations. Selection was assumed to be present but estimated not to be strong enough to overwhelm the effects of drift in such small populations.

---

43. This and other theoretical problems arising from Wright's work were treated from a more comprehensive and mathematically rigorous viewpoint by Gustav Malecot in France (Malecot 1966, 1969; Nagylaki 1989).

44. Lerner and Hazen (1947); Lerner (1958, 117, 223–54); and Snyder (1940, 305) For Wright's views on progress in relation to evolution, see Ruse (1996a, 367–85, 401–6).

In each of the three experiments, the frequencies of a specified mutation (forked, bar, aristopedia, and spineless) competing with its allele were determined for about one hundred lines. The results (including the eventual fixation of one mutation) confirmed the theoretical predictions.[45]

According to James Crow:

> Kerr's experiments were done while he was a postdoc in my lab. I chose the markers, we planned the experiments together, and he carried them out here. Because of his interest in bees we emphasized the X chromosome to mimic a haploid-diploid species. We chose one nearly neutral X-linked gene, one strongly selected X-linked gene, and one autosomal balanced polymorphism. I wasn't sure how to analyze the data and of course thought of Wright. By the time Wright got involved Kerr had moved to Dobzhansky's lab at Columbia. I think Dobzhansky also suggested Wright. Actually, when Sewall and I were talking about these experiments after he came to Madison I found that he had not known where the experiments were done. (Crow, in review of an earlier version of this book, June 2005)

The Kerr-Wright papers were an important contribution to population genetics but did not correct the misunderstanding of Wright's theory by other biologists.[46] Despite the fact that Wright had explicitly included the effects of selection in designing and analyzing the experiments, even his supporters continued to assume that he believed drift to be an *alternative* to natural selection. For those who read the Kerr-Wright papers more carefully, the extremely artificial conditions of this laboratory experiment probably reinforced the criticism that genetic drift, while *theoretically* present in any population, is unlikely to play a significant role in evolutionary processes in nature.

The attacks by Fisher and Ford on Wright's theory of genetic drift, and Wright's attempt to clarify and perhaps revise his position, have been treated in detail by Provine, and the continuing controversy has been analyzed from a philosophical viewpoint by several scholars.[47] Philosophers of biology still argue with each other about whether drift can be unambiguously defined

---

45. Kerr and Wright (1954a, 1954b) and Wright and Kerr (1954).

46. Lewontin (1967, 64) and Provine (1986, 456).

47. Provine (1985, 1986); Hodge (1987, 1992); Skipper (2000, 2002); Millstein (2002, 2005, 2007a, 2007b); and Plutynski (2005, 2007b). Turner (1987) offers some useful insights on the subject.

and distinguished from natural selection and whether they should be considered causes or effects.[48]

In surveying the response to this controversy by other biologists, I found it necessary to distinguish between two hypotheses, both known as genetic drift and/or the Sewall Wright effect:

1. Evolution occurs fastest in medium-size or large populations, divided into several partially isolated smaller populations, in which certain characters first appear in one subgroup as a result of genetic drift and then are incorporated into others by natural selection. In this case, genetic drift has evolutionary significance only *in combination* with natural selection.

2. A nonadaptive character may be explained as the result of genetic drift *rather than* natural selection.

Wright usually meant (1) and rejected the ascription to himself of (2), even by those who (like Motoo Kimura) thought they were supporting Wright's theory.[49] Perhaps his best explanation of (1) was his analogy of Mexican jumping beans on a surface with several hollows of varying depths (Wright 1960). The hollows correspond to adaptive peaks, but gravitational force replaces selective force. The only way the bean can move from a shallow place to a deeper place is by randomly jumping around; gravity (selection pressure) will just keep it where it is. Perhaps if he had used the terminology of the kinetic theory of gases (or more precisely of statistical mechanics) to formulate this physical analogy, with thermal fluctuations playing the role of genetic drift in making possible the transition from a shallow potential well to a deeper one, he could have communicated his idea to Fisher more effectively. But Fisher would still have objected that in a multidimensional space of gene frequencies, there is no reason to think there is even one position that represents an absolute minimum (meaning maximum fitness) for *all* gene ratios (Fisher 1941, 58).

Taken literally, Wright's shifting balance theory is vague and incoherent;[50] it is perhaps best understood as a heuristic or qualitative metaphor (Skipper 2004). But even though "some of its underlying assumptions are viewed as contradictory," Michael Wade and Charles J. Goodnight were

---

48. Brandon (2005); Millstein (2006, 2007a); Plutynski (2007b); and Reisman and Forber (2005).

49. Moore (1953, 228) and Wright (1951, 1955).

50. Fisher (1941, 1953); Coyne et al. (1997); Provine (2001); and Ruse (2004).

able to design an experimental test of the interdemic phase of Wright's theory with favorable results.[51]

Hypothesis (2) was much more popular among biologists, not only because it was simpler, but also because it seemed to solve one of the outstanding difficulties of natural selection theory: the fact that the most prominent visual differences between closely related species seem to have no adaptive value. How then could natural selection explain the differences between them? This objection was made most effectively by O. W. Richards and G. C. Robson and was probably a major reason for the reluctance of some biologists to accept natural selection as an explanation for speciation in the 1930s.[52] Since the theory of Fisher, Haldane, and Wright implied that there must be genetic drift in sufficiently small populations, those who did not understand the various ramifications of the theory assumed that it was legitimate to invoke genetic drift to explain any nonadaptive character.[53] But the result was often unfair to Wright, since his theory lost credibility when it was judged by the outcome of predictions that he himself did not make (and probably would not have made).

The popularity of genetic drift in version (2) was also due in part to the widespread publicity about randomness in quantum physics. If God could play dice with atoms (an assertion that Albert Einstein helped to publicize by denouncing it), why not with genes and chromosomes, which must be made of atoms? A corollary of the unpredictability of atomic motion was the unpredictability of biological phenomena, which was translated into a justification for relaxing the requirement that scientific theories must make testable predictions.[54]

As Wright and others pointed out, he was not the inventor of the *concept* of genetic drift; he just proposed it with a handy name at an opportune moment. Somehow the phrases "accidental multiplication and decline of mutant genes" (Muller) and "genetic-automatic processes" (Dubinin and Romashov) did not catch on. But as Mark B. Adams suggested, the "simultaneous discovery" by Wright and Dubinin in 1931 could reflect broader "intellectual currents of thought" at the time (Adams 1968, 26). One such current, which includes the indeterminism of quantum mechanics, is the neoromantic revival of holism and antireductionism in science in the in-

---

51. Wade and Goodnight (1991, 1015); see also Crow's "perspective" (1991, 973) accompanying that paper and the comprehensive review by Wade (1992).

52. Richards and Robson (1926); Robson (1928); and Robson and Richards (1936).

53. There was also some confusion between characters that are *adaptive* for an organism in a particular environment and those that give an *advantage* to one organism over another; see Millstein (2007b).

54. Scriven (1959) seems to imply this; Villee (1954, 4) makes it more explicit.

terwar period.[55] I return to this issue in chapter 22, but here I should again remind the reader that in this monograph, I follow the selection versus drift controversy only up to 1970. Looking back from 1970, it appears that selection had triumphed, but as those familiar with more recent developments know, drift was revived by Kimura and others in the 1970s and is still a significant part of the modern theory of evolution.

55. Brush (1980); Harrington (1996); and Lawrence and Weisz (1998).

....

4

....

Haldane: A Mathematical-Philosophical Biologist Weighs in

While the fundamental works of Fisher (1930a) and Wright (1931, 1932) seem to have had the greatest effect in persuading biologists that natural selection could account for evolutionary history, the earlier papers of J. B. S. Haldane, beginning in 1924, have been credited with first establishing that natural selection could account for the evolutionary change going on now.[56] According to Sahotra Sarkar (2004), Haldane was the first to propose a true *synthesis* of population genetics with classical genetics, cytology, paleontology, and evolutionary theory (see also Carson 1980). "Fisher's and Wright's theories were retrospective in intent; Haldane's was purely prospective [intended to predict processes in the future] though it allowed retrodiction. To the extent that an *evolutionary theory* should be one that attempts future prediction, only Haldane's work qualifies."[57] In particular,

> Haldane's analysis of gene frequencies in *Biston betularia* (Haldane 1924) gave genetic Darwinism its most famous paradigm case. . . . Haldane . . . proved that the black pepper moth then proliferating was being pumped up by selection pressure in ways that matched data to model, and hence gave empirical support to treating natural selection in terms of changing gene frequencies.[58]

Haldane's 1932 book, *The Causes of Evolution*, addressed the non-mathematical reader with a lively style, providing a large number of biological examples of natural selection. At the same time, he was anxious to

---

56. Sarkar (2004) and Depew and Weber (1995).
57. Sarkar (2004, 1223); he credits the retrospective/prospective distinction to Warren John Ewens (1979, 1990).
58. Depew and Weber (1995, 237–8).

....

take advantage of his own prestige as a mathematician, asserting, "I can write of natural selection with authority because I am one of the three people who know most about its mathematical theory" (Haldane 1932b, 33; see also his earlier technical papers included in the appendix). But his many other activities in science and politics kept him from devoting his full attention to evolutionary theory after 1932.[59] At times he seemed a somewhat lukewarm supporter of the natural selection hypothesis, suggesting that natural selection does not always make an organism more fit and may even drive a species to extinction; he was willing to concede a small Lamarckian effect over long time periods and an orthogenetic "evolutionary momentum" in a few cases.[60] His changing views on the importance of drift will be noted in chapter 18.

Haldane was clearly more interested than Fisher or Wright in philosophical and theological issues. He characterized his own position as materialistic monism and attacked the holistic doctrine of "emergence" as hostile to scientific progress. Some biologists postulated new entities whose properties cannot be predicted from those of its parts. The fallacy of this approach was shown by the example of quantum physics: the properties of the hydrogen atom seemed much more complex than those of its parts (electrons and protons), yet physicists did not abandon the attempt to explain the atom in terms of the properties of its parts. "Wave mechanics thus represents the most serious inroad yet made on the doctrine of emergence, which attempts to set up barriers to the progress of scientific interpretation" and "gives hope to biologists who seek a rational understanding of evolution."[61]

As for the creationists who believe in (what is now called) intelligent design, their doctrine, Haldane asserted, is refuted by two hard facts: first, the extinction of many species; and second, the evolution of parasites that

---

59. Haldane (1932b, 13). Haldane reserves his most emphatic praise for Fisher's result that *"a normally distributed population cannot be in stable equilibrium as a result of selection for the characters normally distributed.* This rather sensational fact vitiates a large number of the arguments which are commonly used both for and against eugenics and Darwinism" (Haldane 1932b, 196). Wright's results are important, Haldane writes, but he underestimates the importance of dominance (Haldane 1932b, 212).

On Haldane's life and personality, see R. W. Clark (1969) and J. M. Smith (1988, 1992). He was sometimes confused with his father, J. S. Haldane, whose views on many subjects were quite different (J. B. S. Haldane 1932a, vi). According to Gordon McOuat and Mary P. Winsor (1995), Haldane took up the study of evolution because of his interest in religion and annoyed some readers by attributing the assertion "Darwinism is dead" to all theologians.

60. Haldane (1932, 12, 23, 108, 136–7).

61. Haldane (1932b, 156–7; 1932a, 262).

have lost their faculties and inflict pain on other animals. Could a morally perfect God have made the tapeworm (Haldane 1932b, 159)? The common statement (still frequently made by creationists in the twenty-first century) that "natural selection cannot account for the evolution of a highly complex character" is wrong, and such "common-sense" conclusions "are often very doubtful"—we need to use mathematics to get correct answers, Haldane believed.[62]

62. Haldane (1932b, 213); see McOuat and Winsor (1995).

5

....

Early Reception of the
Fisher-Haldane-Wright Theory

The works of Fisher, Wright, and Haldane published at the beginning of the 1930s attracted little attention before 1935 except among a small group of their colleagues in Britain and America, most notably E. B. Ford at Oxford.[63] According to Fisher, this was because of the unfortunate influence of William Bateson and his wrongheaded ideas about evolution.[64] Fisher was also annoyed that T. H. Morgan could not see that his own work had made a genetic theory of evolution based on selection of small mutations not only possible but irresistible.[65] Ironically (in view of the supposed incompatibility of evolution and religion), the first positive response came from the Bishop of Birmingham, who asserted that Darwin's theory was now completely confirmed: "Such statistical investigations as those in R. A. Fisher's recent volume are a triumphant vindication of the

63. Ford (1931) and Poulton (1931); see books published in the years 1932–1934 by A. F. Shull and E. Moore and articles by Babcock, L. C. Dunn and Landaur, R. R. Gates, R. Goldschmidt, Harland, and H. F. Osborn. The statement in the text is based on my surveys of *American Naturalist, Genetics,* and *Journal of Genetics.* At the suggestion of Will Provine, I also examined selected publications of the sixteen invited speakers (other than Fisher, Haldane, and Wright) at the Sixth International Congress of Genetics (1932). For a discussion of this congress, see Crow (1992b). According to Bennett (in Fisher 1999, foreword), "Of the 1500 copies printed [of Fisher's *Genetical Theory*], over one-third were sold in the first year. However, sales soon declined and the last copies were not sold until 1947. Many biologists at the time presumably did not read or understand Fisher's book."

64. Fisher, letter to E. B. Poulton, undated (ca. 1930), in Fisher papers at University of Adelaide. I thank Nancy Hall for copies of this and other letters from the Adelaide archive.

65. Fisher, letter to L. Darwin, November 15, 1932, and letter to T. H. Morgan, October 11, 1932, in Fisher papers; see Morgan (1932). Ten years later Fisher was still annoyed about geneticists who "discredited themselves" by opposing Darwinian evolution in the first quarter of the century; he wrote to K. Mather on February 5, 1942, that "the ideas of this period are permanently embalmed in amber in Morgan's mind" (quoted in Bennett 1983, 235–6). On Morgan's views about evolution, see Allen (1968, 1978, 1980); Bowler (1978); Dobzhansky (1959a, 254); and Powell (1987). His eventual acceptance of natural selection is discussed by Muller (1946).

....

potency of natural selection."[66] Also ironically (in view of the well-known reluctance of the French to accept Darwinism), the most substantive discussion of the mathematical theory was in a booklet published in Paris by Philippe L'Héritier (1934).

But the standard argument that natural selection without large (and perhaps *directed*) mutations could not originate new species was still prevalent among biologists through the 1930s.[67] It was only dimly recognized that H. J. Muller's experiments on the production of small mutations by x-rays might provide important evidence on the nature of such mutations.[68] Mathematical derivations were admired but did not convince many readers that they entailed real biological consequences. Evidence from genetics was not convincing to, or even known by, naturalists who considered the field more important than the laboratory.[69]

Gradual Acceptance of Natural Selection

In 1935 we find a few indications that the Fisher-Haldane-Wright (FHW) theory had become part of the discourse of evolutionists about mutations,[70] although their attention to the theory was somewhat distracted by arguments about Fisher's theory of dominance (East, 1935).

Four major publications in 1936 show that views on the causes of evolution were in a state of flux.

A. F. Shull, in his textbook *Evolution*, declared that after years of neglect, "the theory of natural selection is coming back," but he acknowledged that

---

Allen (1979) provides a tabulation of views on evolutionary mechanisms of biologists active in the early twentieth century.

66. Barnes (1931); see more extended comments in his book (1933). In her biography of her father, Joan Fisher Box writes that Fisher gave a copy of his book to Barnes, "an old friend in the Eugenics Society" (Box 1978, 193) and later notes that Barnes was "godfather of his youngest daughter" (Box 1978, 279). According to James Moore, Barnes was Fisher's "old tutor" (Moore 2007, 131). See also Bowler (1998, 2004).

67. Goldschmidt (1934, 171) and Robson and Richards (1936). The prominent botanist Ernest B. Babcock argued that the Fisher-Haldane-Wright theory is inadequate to explain the evolution of new species in the genus *Crepis* because that theory is based on gene mutation. Babcock believed that one must postulate transformation of entire chromosomes, such as the reduction from ten to eight chromosomes, as well as interspecific hybridization and polyploidy (Babcock 1934).

68. Muller (1927) and Crow and Abrahamson (1997). Muller also published a readable comprehensive survey in the popular magazine *Scientific Monthly* (1929). One of the first biology textbooks to mention this work, for which Muller received the Nobel Prize in 1946, was Holman and Robbins (1934, 577).

69. Kingsland (1995) and Mayr (1980c). On the split between experimentalists and naturalists, see the articles by Garland Allen (1979) and Kohler (2002); but Hagen (1982, 1999) questions the validity of this dichotomy.

70. E. R. Dunn (1935); Muller (1935); and Plough and Ives (1935).

traditional and more recent objections still have some validity. It almost seems that we are forced to believe in natural selection because all the other known alternatives are even less credible.[71]

Robson and Richards, in their monograph, *The Variation of Animals in Nature*, also asserted that most of the traditional evidence for natural selection was weak, but they were much less optimistic than Shull about the prospects for rescuing the theory. The mathematical theories of Fisher, Haldane, and Wright, while valuable as a stimulus to future work, were based on unrealistic assumptions and "[did] not in fact provide any proof of the efficacy of selection," other than the alleged absence of any other explanation "for the spread of variants that occur as single or few individuals." But, taken as a whole, "the direct evidence for the occurrence of [n]atural [s]election is very meagre and carries little conviction."[72] According to Stephen Jay Gould and William Provine, this book was very influential and represented the view, widely held at that time, that the differences between species are nonadaptive.[73] Ernst Mayr, on the other hand, said those who held that view were "very much in the minority" (1980a, 132).

A discussion at the Royal Society of London (also in 1936) on the "Present State of the Theory of Natural Selection" began with the following statement by D. M. S. Watson:

> The theory of natural selection is the only explanation of the production of adaptations which is consonant with modern work on heredity, but it suffers from the drawback that by the introduction of related subsidiary hypotheses it becomes capable of giving a theoretical explanation of any conceivable occurrence, and that the scope and indeed the validity of its basal assumption of a selective death-rate determined by a favorable variation have not yet been established. (Watson et al. 1936, 45)

Despite Fisher's vigorous protest (Fisher 1936), Watson's skepticism about natural selection set the tone for the discussion. Gates and MacBride attacked natural selection, favoring instead mutations and Lamarckism, re-

---

71. He asserted that evidence from protective coloration is probably worthless because of McAtee's experiment (examining the contents of stomachs of birds) and other criticisms; the theory can't yet explain how complex organs like the eye arise by small steps, since the early steps give no advantage. Johannsen's negative experiments on pure lines still count against the NSH; since "the observed differences between varieties of a species appear not to be adaptive," we must rely primarily on chance rather than selection to explain their origin" (Shull 1936, 142, 161, 166–83, 205, 207, 212).

72. Robson and Richards (1936, 218, 314, 310).

73. Gould (1983, 89) and Provine (1986, 295).

spectively. Other participants (Timofeeff-Ressovsky, Carpenter, and Haldane) pointed out evidence and arguments in favor of the NSH. Finally, Julian Huxley gave a pro-selection speech to the British Association for the Advancement of Science called "Natural Selection and Evolutionary Progress" (Huxley 1936), the first draft of the manifesto that later became his 1942 book, *Evolution: The Modern Synthesis.*

In 1935, most publications that had discussed natural selection expressed an unfavorable view of it.[74] In 1936 and 1937, the pros and cons were about equally balanced, without counting the publications of Fisher, Haldane, and Dobzhansky.[75] Beginning in 1938 and continuing through the 1940s, a majority of the articles and books that discussed natural selection were favorable.[76]

During the 1940s, the most prominent opponent of the NSH was the respected German biologist Richard Goldschmidt. Goldschmidt accepted natural selection as an adequate explanation for microevolution but argued that it could not explain macroevolution—in particular, the formation of new species. For this, he invoked his own theory of macromutations, using the notorious phrase "hopeful monsters" (Goldschmidt 1940, 390). This idea may have been attractive to a few botanists, since plants occasionally form new species in one generation by polyploidy (combining chromosome sets), as noted by Weatherwax (1947, 391–2) and later by Mayr (1980a, 130). But E. B. Babcock, Goldschmidt's colleague at Berkeley and a leader in research on polyploidy, firmly rejected the saltationist view (Smocovitis 2008), and Mayr called polyploidy "the minority process" in plants.

Goldschmidt's theory became the punching bag for the modern synthesis (Gould 2002, 68). As he realized, "all theories of evolution tend to reflect

74. Books by Nordenskiold and by Whitefield and Wood; articles by Balfour, Blum, Böker, McAtee, Nilsson, Robb, Wheeler. Articles expressing favorable views were those by E. R. Dunn and Sumner.

75. *Pro*-Natural Selection: 1936 book by Shull; articles by Carpenter, East, Gordon, Timofeff-Ressovsky, Turrill; 1937 book by Sturtevant and Beadle; articles by Ford, Just, Kemp.

Anti-Natural Selection: 1936 books by Mavor, Mercier, Robson and Richards; articles by MacBride, Watson; 1937 books by McKerrow, Scott.

76. I give the detailed list for only one more year. *Pro*-Natural Selection: 1938 books by Allee, Dendy, Ford, Shull, Young et al.; 1938 articles by Carr-Saunders, Elton, Gates, Harlan and Martini, Muller, Quayle, Shull, Sturtevant.

Anti-Natural Selection: 1938 book by Bradley. The textbook by Young, Stebbins, and Hylander, counted as "pro" in the above list, is not enthusiastically so and is willing to consider the possibility of orthogenesis. See tables 2 and 3 for publications after 1940.

I am assuming that the periodicals and books I have been able to look at are representative at least of those published in the United States and Britain and that a larger sample would not change the qualitative trend described in the text.

the scientific trends of their time" (Goldschmidt 1940, 397); unfortunately for him, his own theory was a reflection of the holistic, neoromantic biology of the 1930s, which was about to be displaced by the mechanistic reductionistic biology of the 1940s and 1950s.[77]

Beginning in 1942 with the publication of Julian Huxley's *Evolution: The Modern Synthesis* and Ernst Mayr's *Systematics and the Origin of Species*, there was a stream of books and articles by leading biologists, articulating and strongly supporting the evolutionary synthesis. It is now time to ask, what were the reasons for this groundswell of support, and to what extent was it focused on a single coherent theory such as the natural selection hypothesis?

---

77. On Goldschmidt's life, work, and reputation, see Dietrich (1995); Schmitt (2000); and Barash (2003). The reception of his macromutation theory is discussed by Gould (2002, 453ff). Stebbins (1969) declared that Goldschmidt's theory was killed by the new molecular biology based on the Watson-Crick double helix model of DNA. But even before that, leaders in paleontology and evolutionary biology had rejected macromutation as a significant factor in evolution: see Babcock et al. (1942); Huxley (1942); Sumner (1942); White (1945); Allee et al. (1949); Patterson and Stone (1952); Watson (1952); Mather (1953); and Sheppard (1954). Mayr, in response to Dietrich (1995), wrote that Goldschmidt's (1940) book was reviewed gently because everybody felt sorry for him—a few years earlier he was "the pope of German biology ... now in exile and had to make his living by giving undergraduate courses" (1997, 33).

············

6

Dobzhansky:
The Faraday of Biology?

In Russia, where Dobzhansky was born and educated, there was much more support for Darwin's theory of natural selection and much less separation between genetics and natural history than in Britain and America.[78] When he came to the United States in 1927 to work with Morgan's group, first at Columbia and then at the California Institute of Technology, Dobzhansky enhanced his already-substantial knowledge of genetics and used it to continue his field research in entomology (Dobzhansky 1933). His famous series of papers on the genetics of natural populations established a strong link between the established knowledge about *Drosophila* in the laboratory and new knowledge about *Drosophila* in the wild.[79] According to Francisco Ayala, "The single most important empirical fact established and argued by Dobzhansky is the ubiquity of genetic variation" (Ayala 1976, 5).

Dobzhansky's research in the mid-1930s provided strong indirect support for the NSH by showing that speciation may proceed in small steps. He became interested in Donald Lancefield's discovery that a newly discovered strain of *Drosophila*, later called *D. pseudoobscura*, produced sterile males when crossed with other strains, but also produced some fertile females. This "may represent an earlier step in the evolutionary process" as the first step in the origin of a new species (Lancefield 1929, 288). Dobzhansky was inspired by the possibility of creating a new species in the laboratory, a feat previously regarded as impossible.[80] He proceeded by studying the

78. On the nineteenth-century reception of evolutionary ideas in Russia, see Vucinich (1974); on Dobzhansky's own Russian background and his work in Morgan's lab, see Adams (1994) and Allen (1994). Ayala (1990) gives a comprehensive account of his life and work.

79. Dobzhansky (1981); Levine (1995); and Ayala and Fitch (1997).

80. Later he became aware of the creation by G. D. Karpechencko in the 1920s of a new species, *Raphanobrassica*, made by crossing broccoli with radish, and he gave an extensive discussion of it (Dobzhansky 1941b, 233–5, 273, 325).

genetic basis of hybrid sterility (Dobzhansky 1936). He argued that two strains of *D. pseudoobscura* were in fact different species because they were reproductively isolated; one of them was later renamed *D. persimilis*.

His work suggested that speciation was not necessarily a sudden discontinuous process involving a macromutation, thereby challenging the views of those evolutionists who still followed de Vries and preemptively undermining Richard Goldschmidt's theory. Instead, he postulated that there is no qualitative difference between microevolution and macroevolution as a pragmatic methodological assumption[81] and then gathered empirical evidence for this postulate; thus, research on small changes at the level of genes and chromosomes is directly relevant to the general problem of the origin of species. Speciation is not a discrete event but a process that occurs over several generations; a species might be considered what mathematicians were later to call a "fuzzy set."[82] Wright, in his review of Goldschmidt's book, agreed with Dobzhansky that we have examples of every conceivable intermediate step; therefore, there is no justification for assuming the "'bridgeless gap' between species which is crucial" for Goldschmidt's theory (Wright 1941, 166).

Dobzhansky initially relied on Sewall Wright to tell him the conclusions of mathematical population genetics and was influenced to favor Wright's views on the importance of genetic drift (Epling and Dobzhansky 1942), but his own results led him to change those views in the 1940s.[83] Thus, Dobzhansky's career in America illustrates in microcosm the shift toward selectionism in the 1940s and 1950s.

The publication of Dobzhansky's *Genetics and the Origin of Species* in 1937 may be taken as the first announcement of the evolutionary synthesis. Dobzhansky presented a comprehensive review of the biological evidence on mutations, variation, selection, mechanisms for speciation (including polyploidy and isolation), patterns of evolution, and species. He called attention to Muller's x-ray work, which he called the first conclusive evidence for an external cause of mutations. He mentioned, and emphasized in the second edition (Dobzhansky 1941b), the "rediscovery of the giant chromosomes in the larval salivary gland of flies by Heitz and Bauer (1933) and the application of these chromosomes as a tool of genetic research by T. S. Painter (1934)," which "enormously facilitated the comparison of the gene

81. Dobzhansky (1937, 12); Powell (1995, 74); and Orr (1996).

82. Willermet and Hill (1997). For debates about the definition of species, see Dobzhansky (1935); Emerson (1945); Krementsov (1994); Mayr (1948, 1949a); Winsor (2000); and Reydon (2005).

83. Dobzhansky (1943, 1971); see the editorial note on page 303 preceding the reprint of this paper in Dobzhansky (1981, 305–28); Provine (1986, 389–90); and Lewontin et al. (2001).

arrangements in different strains."[84] This tool provided the basis for Dob-zhansky's research in the 1940s on natural populations of *Drosophila*.

Genetics and the Origin of Species went through three editions under that title (Dobzhansky 1937, 1941b, 1951), and a further revision was published under the title *Genetics of the Evolutionary Process* (1970). These are gener-ally considered to have been very influential in persuading biologists to adopt the evolutionary synthesis.[85] One reason for this influence, accord-ing to Leah Ceccarelli (2001), is the rhetorical approach its author used. Rather than claim that one field of science should be reduced to another more fundamental field (as did E. O. Wilson in *Sociobiology* [1975], another book discussed by Ceccarelli), Dobzhansky made it clear that knowledge and research methods from several fields must be integrated in order to understand how evolution works. In particular, geneticists and naturalists would have to cooperate as they had not done in the past (at least in the Anglo-American scientific community). Before 1937, naturalists thought geneticists still believed in de Vries's saltation theory, while geneticists did not appreciate the great amount of variation in natural populations; *Genetics and the Origin of Species* helped both groups to overcome their misconcep-tions and recognize how they could profit from each other's work (Cec-carelli 2001, 28). At the same time, as Lindley Darden has pointed out, the relations between fields in the evolutionary synthesis were not symmetrical, because the synthetic theory had to deal with three levels in a hierarchy: genes and chromosomes, populations, and species. "Since genetics can pro-ceed without taking into account the populational level, it retains more relative independence than do the other fields in the synthesis"; popula-tion genetics depends on "new findings of mutational processes in genetics," and population processes "can continue whether or not isolation [producing new species] has occurred."[86]

---

84. Dobzhansky (1941b, 117). According to M. J. D. White, giant chromosomes in other dipterous flies were discovered as early as 1880. Progress in genetics might have been much faster if *Drosophila* researchers had been familiar with "the work of Balbiani, Carnoy, Alverdes and oth-ers on the salivary chromosomes of *Chironomus*" and had looked at salivary glands twenty years earlier (White 1945, vii, 3). While the "rediscovery" came too late to play an important role in persuading biologists to accept Morgan's chromosome theory of heredity (Brush 2002), it did come just in time to change Dobzhansky's views on the relative importance of natural selection and isolation.

As Muller noted in his own survey of research on mutations, Dobzhansky was one of those who confirmed the effects of x-rays (Muller 1929, 495).

85. Gregory (1946); Gould (1982); Provine (1994); Krimbas (1995); Ceccarelli (2001, 26); and Dunn (1961 reminiscences quoted by Provine [1994, 100–1]).

86. Darden (1986, 121, 122). She also notes that the asymmetrical hierarchical relations that were assumed in 1937 tended to break down as a result of later research: "Selection [on the pop-ulation level] may influence the evolution of gene mutation mechanisms" (Darden 1986, 122).

In the first chapter of the third edition of *Genetics and the Origin of Species*, Dobzhansky distinguished two approaches to evolutionary problems: (1) "unraveling and describing the actual course which the evolutionary process took in the history of the earth" (this is phylogeny); and (2) "studies on the mechanisms that bring about evolution, causal rather than historical problems, phenomena that can be studied experimentally rather than events which happened in the past" (Dobzhansky 1951, 11). This is an elaboration of a brief statement in the 1941 edition (6), where he attributes the identification of these two approaches as "generalizing" and "exact induction" to M. Hartmann (1933). In the spirit of the evolutionary synthesis, the two approaches are complementary rather than antagonistic.[87]

For other recent discussion of the scientific issues raised by Dobzhansky's book, see Lewontin (1997).

87. The two approaches correspond to the prospective and retrospective categories used by Sarkar to distinguish the work of Haldane from that of Fisher and Wright; see note 57. Dobzhansky's use of the distinction is discussed by Ayala (1977) and Prout (1995).

7

....

Evidence for Natural Selection before 1941

For our purposes, the most relevant part of *Genetics and the Origin of Species* is the survey of evidence for natural selection, much of which could provide reasons for biologists to support the more specific natural selection hypothesis based on small mutations. I have used the more extensive list found in Dobzhansky's second edition,[88] in which he organized the evidence into six categories.

Laboratory Experiments

In the *Origin of Species*, Charles Darwin discussed the "remarkable fact," discovered by Thomas Vernon Wollaston a few years earlier, that many of the beetles inhabiting the Madeira Islands in the Atlantic Ocean are wingless, unlike related beetles elsewhere. Darwin suggested that

> the wingless condition . . . is mainly due to the action of natural selection, but combined probably with disuse. For during thousands of successive generations each individual beetle which flew least, either from its wings having [been] ever so little less perfectly developed or from indolent habit, will have had the best chance of surviving from not being blown out to sea; and, on the other hand, those beetles which most readily took to flight will oftenest have been blown to sea and thus have been destroyed.[89]

This suggestion, elaborated in Darwin's "big book" on natural selection (Darwin 1975, 291), is equivalent to a prediction that if you start with a population of beetles with varying wing sizes and put them in an environment where those that have the largest wings are most likely to be blown

---

88. Dobzhansky (1941b, 188–214).

89. Darwin (1859 [and later editions], chapter 5 on "effects of the increased use and disuse of parts"). For correspondence and information relating to this suggestion, see Darwin (1989, 267–74, 279–80, 283–5). I thank Sandra Herbert for this reference.

....

away by strong winds, then (assuming wing size is an inherited character) after many generations, the population will consist of a much larger proportion of beetles with small or absent wings.

In the following decades, Darwin's discussion of the wingless flies of Madeira was often given as an example of how an observation could be explained by natural selection: the insects with wings were blown away; the others remained (Thomson and Geddes 1931). But it was not clear whether selection acted on large or small mutations, or even whether wind was the most important selective factor. According to the (Popperian) scientific method, we should do an experiment in which the intensity of a mechanically produced "wind" is varied while other factors are kept constant; our hypothesis predicts that over the course of many generations, the proportion of wingless flies should increase if the wind is strong but not if it is weak or absent.

A critic unfamiliar with modern biology might scoff at this proposal, because "everyone knows" that evolution is such a slow process that it can't be directly observed within a human lifetime, much less in a laboratory experiment lasting only a few weeks or months. But the critic has ignored the fact that geneticists in the early twentieth century developed a "model organism," the *Drosophila* fly, which reproduces so rapidly that such experiments can be done in the laboratory in real time. By 1930, any good geneticist could do *Drosophila* research and even assign the classical experiments as practical exercises for students.[90]

In 1937, in France—not usually considered a favorable environment for research on natural selection at that time[91]—the test was done. Philippe L'Héritier, Yvette Neefs, and Georges Teissier carried out an explicit test of Darwin's prediction. They used a "population cage," previously developed by L'Héritier and Teissier (1934), which allowed one to study an isolated population under controlled conditions.[92]

Nous nous sommes proposé d'etudier expérimentalement le problème posé par Darwin et nous avons utilisé pour cela Drosophila melanogaster. *Nous avons réalisé les conditions que Darwin suppose à l'origine de l'évolution des Insects insulaires en constituant une population mixte d'individus normaux ailés et*

90. Kohler (1994) and Brush (2002, 519).

91. Lamarckism dominated French biology until after World War II, according to Buican (1983, 1989) and Grimoult (2000). L'Héritier (1981) attributes this to Catholics who think it is more moral than Darwinism.

92. L'Héritier (1981); Grimoult (2000, 151–63); and Gayon and Veuille (2001). L'Heritier spent the academic year 1931–1932 in the United States (mostly at Iowa State University), where

d'individus portant le caractère récessif vestigial (vg) *qui se manifeste par une atrophie des ailes devenues tout à fait impropres au vol.* (L'Héritier et al. 1937, 907–8)

[We decided to study experimentally the problem posed by Darwin and have, for this purpose, used *Drosophila melanogaster.* We have realized the conditions that Darwin assumed to exist originally: a mixed population of some insects with normal wings and others having the recessive character *vestigial* (*vg*), which is manifested by an atrophy of the wings, so that they have become inadequate for flight.]

And indeed, in a trial of less than two months, they were able to confirm the results predicted from Darwin's hypothesis ("*exactement celle que permettait de prévoir l'hypothèse de Darwin*"). Although L'Héritier himself later characterized this as an "*experience amusante*" (1981, 338), and Haldane called it a "rather trivial but still illuminating experiment" (1939, 127), C. H. Waddington identified it as a confirmed prediction and wrote that it is "an example of a type of work which is of considerable importance for the experimental study of evolution but which has been surprisingly little taken up" (1939, 302).

Dobzhansky, who gave this as his first example of natural selection in the laboratory, did not mention Darwin's role, nor did he use the language of hypothetico-deductivism and prediction-testing.[93] Instead, he quickly moved on to the next category.

Historical Changes in the Composition of Populations

Dobzhansky conceded that in the case of major changes, such as the evolution of the horse or the anthropoid ape, the generic critic of evolution

---

he learned about the work of Fisher and Wright. He thought of the idea of the population cage while strolling on an American beach. According to Gayon and Veuille, the Darwinian experiment "provided its authors with lasting fame in the polemical context of French studies on evolution. . . . This experiment was an effective reply to adversaries of Darwinism. It as much seemed an artist's performance as a scientific investigation, its purpose probably being to ridicule the numerous French opponents of 'transformism.' Denying natural selection was like flying in the face of the wind" (Gayon and Veuille 2001, 89).

The experimental technique was rather sloppy by modern standards; while the wind strength was "controlled," there was apparently no control group (a similar population not exposed to wind). I thank David A. O'Brochta for a critique of the experiment (personal communication with author, March 4, 2005).

93. Dobzhansky (1941b, 189). In the 1970 revision of *Genetics*, Dobzhansky omits this example entirely. I have not found any statements in the literature up to that time other than Waddington's (quoted in the text), claiming that the experiment was a confirmation of a prediction.

(whom I invoked above) would be correct: evolution was supposed to be so slow that no changes could be noticed within a human lifetime in wild species. So, Darwin and his followers had to use the results of *human* selection and breeding as evidence for the theory of *natural* selection: "The relatively rapid changes brought about in domestic animals and cultivated by artificial selection were considered a model of the evolution in the wild state, rather than evolution as such" (Dobzhansky 1941b, 190).

At the beginning of the twentieth century, fumigation attempts with hydrocyanic gas to eradicate red scale insects that attack citrus trees in California were initially successful, but after a few years it was observed that fumigation no longer worked. Experiments by H. J. Quayle (1938) showed that there were two races of red scale—one of which is resistant to the gas, the other not. By eliminating the nonresistant insects, the fumigation allowed the resistant race to take over the population. Dobzhansky called this "probably the best proof of the effectiveness of natural selection yet obtained" (1937, 161).

But where did the hydrocyanic gas-resistant insects come from? Dobzhansky suggested that they resulted from mutations, rather than from being "introduced from elsewhere," but he considered this question "largely academic"—a rather inappropriate pejorative for a professor to use! In a later revision of the book, he seemed unsure whether the mutations "are present in pest populations before the insecticide is applied, or arise after the application. This question is usually insoluble, and it is of no particular importance anyway" (Dobzhansky 1970, 215). The context for these statements is the shift in the views of many evolutionists toward the position (supported by Dobzhansky's own experiments) that there is so much hidden genetic variability in any present-day population that new mutations are not really needed to drive evolution (Stebbins 1970, 182).

After mentioning other examples of the evolution of agricultural pests (citricola scale, codling moth, and stem rust of wheat), Dobzhansky discussed "industrial melanism": the appearance of darker forms of moths in areas near industrial cities in England and other European countries. He seemed to accept E. B. Ford's explanation (1937, 1940b) that

> melanics are superior to the light-colored types in vigor, and ... their spread in populations is normally prevented because they are not protectively colored. In industrial areas this disability is removed by the general darkening of the landscape.[94]

94. Dobzhansky (1941b, 196); similar statement in Dobzhansky (1951, 132–3).

Experimental Study of Adaptation in Plants

Dobzhansky noted the earlier work on plant races adapted to a specific habitat by "the Swedish botanist and geneticist Turesson ... who must be credited with having placed it on a truly scientific basis" (Dobzhansky 1941b, 197). But he gave much more space to the recent research of Jens Clausen, David D. Keck, and William M. Hiesey in California (Clausen et al. 1940). In order to show that the differences in phenotype of plants of the same species growing in different habitats (called "ecotypes") are hereditary (and therefore due to natural selection) rather than environmental (directly responding to that habitat), they used

> the method of reciprocal transplants[:] ... Representatives of the population of the habitat A are transplanted in the habitat B, and those from B are planted in A. Comparison of the native and transplanted strains permits the observer to discriminate between the changes induced directly by the environment and those intrinsic to the strains themselves. (Dobzhansky 1941b, 197–8)

For example, there are three ecotypes of the cinquefoil (five-petaled rose, *Potentilla glandulosa*): (1) the "coastal ecotype," growing in the Coast Ranges of California, a region which "has a rainless summer and an equable climate permitting the plants to grow almost throughout the year"; and (2) ecotypes in the foothills of the Sierra Nevada Mountains—one in the dry slopes, the other in the meadows. Although the coastal ecotype does not survive very long when transplanted to the much harsher conditions of the foothills, in general, the ecotypes tend to carry over their most distinctive characteristics when transplanted to a different habitat, with minor modifications.[95]

95. This is my rough summary of the rather confusing discussion given by Dobzhansky (1941b, 197–9). This and subsequent work of Clausen et al. are described in more detail in Dobzhansky (1970, 292–5); see the discussion by Smocovitis (1988, 214–33). Dobzhansky also discussed the experiments of W. Sukatschew (1928) on the competition of several strains of the dandelion (*Taraxacum officinale*) grown near Leningrad and brought from other places (Archangel, Vologda in the north, and Askania-Nova in the south). When planted together at high densities, the Archangel and Vologda strains did better than the locally grown strains, but did not when planted at lower densities. (The one from the southern region did not survive in either case.) The conclusion was that survival depends not just on environmental conditions but also on additional factors such as the presence of competing plants. This example was omitted from the fourth edition (Dobzhansky 1970).

Experimental Studies of Adaptation in Animals

Dobzhansky (1941b, 203–5) discussed the experiments of N. W. Timofeeff-Ressovsky in the 1930s on the survival values of strains of *Drosophila funebris* from twenty-four geographical areas in Europe at three different temperatures. Each was forced to compete against a standard strain of *Drosophila melanogaster* by putting 150 eggs of one strain of *D. funebris* with 150 eggs of *D. melanogaster* in a culture bottle and counting the number that hatched. Presumably the (unstated) hypothesis was that since "*D. melanogaster* is native to the Tropics while *D. funebris* occurs in the Temperate Zone," strains of *D. funebris* from colder regions (Russia, Sweden) would survive relatively better at the lowest temperature (15°C) than at the highest (29°C) (Dobzhansky, 1941b, 203–204). The numbers seem to confirm this hypothesis, although Dobzhansky made no attempt to give a statistical analysis of their significance, and again he did not use the language of prediction-testing.

Dobzhansky also reported the results of several other studies in this chapter—Dice on the pocket mouse, Goldschmidt on the gypsy moth, and Fox on crustaceans—but these studies seem to consist mainly of qualitative effects of environmental variations with ad hoc explanations, rather than systematic experiments. Most of this material was gone or used for other purposes in the 1970 edition.

Regularities in Geographical Variation

Here, the emphasis is on the similarities of different species in the same area and on the similar variations found as one goes from one area to another. "This phenomenon is a counterpart on the supraspecies level of the ecotypic differentiation with species. Zoology is far outdistanced by botany in the formulation of regularities of this kind." For example,

> Gloger's rule (frequently referred to in the American literature as Allen's rule) states that in mammals and birds, races inhabiting warm and humid regions have more melanin pigmentation than races of the same species in cooler and drier regions; arid desert regions are characterized by accumulation of yellow and reddish-brown phaeomelanin pigmentation. (Dobzhansky 1941b, 207)

What do these rules have to do with natural selection?

> Strange as it may seem, the correlations between race formation and environment revealed by the "rules" . . . were repeatedly quoted as arguments

against the natural selection theory. This amazing confusion of thought was due in the past to the almost universal acceptance among biologists of the belief in inheritance of acquired characteristics. . . . To suppose that geographical races were originally modifications that have subsequently been fixed by heredity is contrary to the whole sum of our knowledge. . . . Therefore, the regularities observed in the process of geographical race formation cannot be due to direct effects of the secular environment, albeit the results of this process can in part be imitated by some phenotypic modifications. (Dobzhansky 1941b, 208, 210)

In other words, all of this empirical evidence was previously regarded as support for Lamarckism, but since we now know Lamarckism is false, it has to be regarded instead as support for the only plausible alternative that accounts for correlations between adaptations and environment: natural selection.

Protective and Warning Coloration

As for protective and warning coloration, including mimicry—evidence for natural selection going back to the nineteenth century (Kimler 1983a, 1983b)—Dobzhansky acknowledged that much of it is "uncritical and valueless speculation . . . bringing only disrepute to the whole theory. . . . Some of the alleged protections and warnings have been shown to be armchair protection and museum mimicry." But despite heavy criticism in the 1930s by Heikertinger and McAtee, the theory was revived and became quite respectable, as shown by the treatise of H. B. Cott (1940). There was still not much careful experimental work; Dobzhansky mentioned only the research of Francis Sumner (1935) showing that "fishes whose color contrasts with their surroundings are caught by predators more easily than those with harmonizing colors." As for McAtee's "refutation" (showing that the stomachs of birds contain supposedly poisonous insects in the same proportion as their relative abundance in their environment), Dobzhansky argued that one should not assume that "only an absolute immunity from predators can make natural selection effective." All that was needed, according to the modern theories of mathematical population genetics, is a selective advantage of one part in one thousand to cause a significant change in the composition of the species (Dobzhansky 1941b, 212–4).

Readers familiar with the history of evolutionary biology will recognize that Dobzhansky's survey of the evidence was very incomplete, but my goal is not to evaluate or supplement Dobzhansky as a historian; rather, it is to suggest how the presentation of the evidence in his very influential treatise may have set the agenda for others.

. . . .

8

....

Huxley: A New Synthesis
Is Proclaimed

Julian Huxley, according to E. B. Ford, was an inspirational mentor
and "the most powerful force in developing the selectionist attitude at
Oxford in the 1920s."[96] Huxley himself recalled:

> One of the greatest changes in biological outlook during my active life
> has been the re-erection of *natural selection as the main and perhaps the sole
> agency of significant evolutionary change and adjustment....* Perhaps owing
> to a certain familiarity with natural history, I clung firmly to a belief in
> the principle of natural selection and its efficacy during the period up to
> the early or middle 1920s, when it was being neglected or attacked by
> most of the biologists who thought of themselves as advanced, and when
> even eminent geneticists could seriously assert that without natural selec-
> tion all existing types could have come into being, and a vast number of
> others as well![97]

When T. H. Morgan visited Oxford some time in the early 1920s, Huxley
"showed him some of the wonderful series illustrating mimicry in butterflies which Poulton had amassed.... 'This is extraordinary!' he said, with
his vivid enthusiasm; and then, 'I just didn't know that things like this
existed.'"[98]

Huxley's edited book, *The New Systematics* (1940), helped to publicize
the work of specialists such as Darlington, Ford, and Timofeeff-Ressovsky.
In his introduction he wrote:

96. Ford (1980, 337). "J. S. Huxley ... played a central role in creating a neo-Darwinian
synthetic view of evolution in England" (Provine 1980, 331). On his relation to systematics, see
Hagen (1982, 1984). On his worldview, see Greene (1990).
97. Huxley (1951, 593; emphasis added); cf. Huxley (1926).
98. Huxley (1951, 593); also Dronamraju (1993, 237).

....

There is still a widespread reluctance, especially among some of the younger experimental biologists, to recognize the prevalence of adaptation and the power of selection. This is doubtless in large part a natural reaction against the facile arm-chair reasoning of a certain school of evolutionists.

While endorsing the importance of natural selection, he also argued that (geographical) "isolation is the essential factor in bringing about taxonomic divergence" and stated that "where isolation is relatively or quite complete and the isolated population small, the Sewall Wright effect (1931, 1939) will produce a certain degree of random, nonadaptive change." But "isolation (apart from the accidental fixing of new combinations of genes already present) is powerless to effect differentiation without mutation, and, in most cases, without selection" (Huxley 1940, 12).

Huxley introduced the term "synthesis" and popularized his own version of it with the title and readable style of his influential 1942 book, *Evolution: The Modern Synthesis*. He praised the work of Fisher, who

has radically transformed our outlook on the subject, notably by pointing out how the effect of a mutation can be altered by new combinations and mutations of other genes. Any originality which this book may possess lies partly in its attempting to generalize this idea still further, by stressing the fact that a study of the effects of genes during development is as essential for an understanding of evolution as are the study of mutation and that of selection.[99]

He asserted that thanks to the mathematical work of Fisher, Haldane, and Wright, we can now make "quantitative prophecies with much greater fullness than was possible to Darwin" (Huxley 1942, 21). But it is difficult to find any such prophecies in his book, whose pages are remarkably free of numerical data.[100] Huxley's book is nevertheless a comprehensive qualitative survey of the arguments and evidence for the selectionist view. As such, it is not a synthesis of different *theories* but of *observations* in support of one theory (Waters and van Helden 1993, 24–25). Will Provine (1988, 1992) argues that it is a *constriction* that eliminated most of the factors previously believed to affect evolution.

99. Huxley (1942, 8).
100. Ruse (1996a, 329) writes that Huxley, by his own admission (in a 1921 letter to A. Hardy), "did not understand enough mathematics, physics, and chemistry" to study physiology at Oxford. See also R. W. Clark (1968) and Waters and van Helden (1993).

Huxley noted that the (Lamarckian) induction of melanism (as claimed by J. W. H. Harrison) had not yet been disproved, so the validity of the natural selection hypothesis, in this and other cases, must be established by "the convergence of a number of separate lines of evidence, each partial and indirect or incomplete, but severally cumulative and demonstrative. . . . These various lines of evidence all converge to support a neo-[M]endelian view, some showing that small mutations occur, others that selection is active, that some mutations are potentially beneficial, that through selection of the gene-complex, mutations can be adjusted to the needs of the organism, and that adaptations are genetically determined and vary in type and accuracy with direction and intensity of selection" (Huxley 1942, 116, 122–3).

But what kind of evidence did Huxley himself consider most persuasive for a biologist who was already familiar with the older evidence for natural selection but reluctant to accept it as the *dominant* process in evolution? Huxley explicitly answers this question. First, the mathematical work of Fisher and Haldane, combined with modern genetics, definitively refutes the three major alternatives to the NSH. Small mutations are much more likely to produce adaptive improvements than large ones, overwhelmingly so when (as now seems probable) "the harmonious adjustment of many independently varying characters is required"—hence the de Vries large mutation theory is untenable. Similarly, orthogenesis and Lamarckian effects, even if present, cannot be important in evolution as long as selection operates (Huxley 1942, 123). But

perhaps the most important single concept of recent years is that of the adjustment of mutations through changes in the gene-complex. Before this had been developed by R. A. Fisher and his followers, notably E. B. Ford, the effect of a mutation was assumed to be constant. A given mutation, we may say, made an offer to the germplasm of the species, which had to be accepted or declined as it stood. . . . To-day we are able to look at the matter in a wholly different way . . . the offer made by a mutation may be merely a preliminary proposal, subject to negotiation. Biologically, this negotiation is effected in the first instance by recombination and secondarily by mutation in the residual gene-complex. (Huxley 1942, 124)

Huxley also recognized "the Sewall Wright phenomenon of drift in small populations" (Huxley 1942, 59). This effect is involved in speciation, and Huxley called it

. . . .

one of the most important results of mathematical analysis applied to the facts of neo-mendelism. It gives accident as well as adaptation a place in evolution, and at one stroke explains many facts which puzzled earlier selectionists, notably the much greater degree of divergence shown by island than mainland forms, by forms in isolated lakes than in continuous river-systems.[101]

In later publications, Huxley became increasingly skeptical about the evolutionary significance of genetic drift. In 1945, commenting on G. G. Simpson's use of Wright's theory to explain gaps in the fossil record, he wrote, "Geneticists may doubt if such populations could persist for the millions of years needed for non-adaptive evolution" (Huxley 1945a, 4). But he was still willing to give drift a role in speciation (Huxley 1946, 5). In 1951 he asserted that the establishment of nonadaptive characters by drift in small populations is much less frequent than Wright thought; Ford, Muller, and others had shown that "truly non-adaptive genes and gene combinations are exceedingly rare" (Huxley 1951, 597). He was not asserting that Wright's theory made incorrect predictions, but rather that the theory was not relevant because adaptation is operating. (Wright would say that his theory does apply in this case if the population is small enough.)

In 1952 Huxley published an article in the American pictorial magazine *Life*, featuring a sensational example of natural selection: the crab *Dorippe japonica*, which bears what looks like the face of a Samurai warrior. According to a Japanese legend, after the Heike warriors were defeated by the Genji in 1155 AD, they "committed mass suicide by throwing themselves into the sea." Later, fishermen noticed crabs that bore a slight resemblance to the face of a warrior and threw them back, since they did not want to eat the supposed reincarnation of a Heike warrior. As a result, the crabs that looked more like a warrior's face were more likely to survive and reproduce, so the resemblance became nearly perfect in later generations. Huxley argued that such mimicry is common; this one just happens to be "more curious and surprising than others."[102]

I find it also curious and surprising that Huxley's example of the Heike crab was not used in any of the books I examined from the period 1953–

101. Huxley (1942, 155, 186, 194, 197, 208, 229, 232, 242, 259–60, 265, 326, 363 [quotation on 200]). See Muller (1929) for a good explanation of this point. Olby (1992, 70) notes that Ford objected to Huxley's inclusion of drift.

102. Huxley (1957, 141). The *Life* article is reprinted therein.

1970. Perhaps after skeptical biologists had become fed up with the plethora of "just so" stories invented by earlier Darwinians, they thought this one was just too good to be true.[103] Or perhaps it was excluded because it was considered an example of artificial rather than natural selection.

103. But Carl Sagan was not reluctant to use the Heike crab as an example of natural selection in his popular television series, *Cosmos*; see Sagan (1980, 25–26) and Smocovitis (1996, 164–5, note 202). Waddington (1969, 113) pointed out that in a 1927 textbook, Haldane and Huxley used as a frontispiece illustrations of fish whose heads looked remarkably like those of the authors, but Waddington did not make an explicit connection with the Heike crab.

9

Mayr: Systematics and the Founder Principle

Modern evolutionary genetics is a highly mathematical subject, yet most biologists in the twentieth century disliked mathematics and resisted using it. How did the subject become established? One might conjecture that the simultaneous emergence of three mathematical biologists—Fisher, Haldane, and Wright—was a very unlikely event, which happened to occur in the early twentieth century. These three founders, although geographically separated, were able to interact with each other thanks to modern communications and transportation technology and thus were allowed to start a new specialty within the discipline of biology. That specialty might have died out if not for the fact that Fisher and Wright were able to team up with two biologists who were not expert mathematicians but were very good at doing empirical work and publicizing their results: E. B. Ford at Oxford and Theodosius Dobzhansky at Columbia, respectively. Those collaborators attracted bright students who carried on their research programs and thereby sustained the new specialty. (Haldane's influence was somewhat more diffuse, in part because he worked on several other topics as well as evolutionary theory, but he did inspire another theorist, John Maynard Smith; see Smith [1988, 126–9].)

This scenario conveniently illustrates what Ernst Mayr called the founder principle. It resembles a special case of random genetic drift, in which a new species is formed by a very small and very nonrepresentative subset of a larger population.

Mayr is known as one of the leaders of the evolutionary synthesis. His book, *Systematics and the Origin of Species*, published in 1942 with an introductory endorsement by Dobzhansky, formulated the synthesis from the viewpoint of taxonomy and ornithology.[104] In 1963 Huxley called him

---

104. Mayr (1942). Eldredge gives a detailed summary and analysis of the book (1985, 44–59).

"undoubtedly the best all-round evolutionary geneticist we now have in the world" (Dronamraju 1993, 236). At the time of his death in 2005 at age 100, he was "widely considered the father of modern evolutionary biology."[105] As a distinguished biologist-turned-historian, he had considerable influence not only on evolutionary theory but also on how its development was to be retrospectively interpreted.[106]

In the opening pages of *Systematics*, Mayr welcomed the new "mutual understanding between geneticists and systematists," exemplified by "Rensch and Kinsey among the taxonomists, Timofeeff-Ressovsky and Dobzhansky among the geneticists, and Huxley and Diver among the general biologists." As for the mathematicians, he mentioned Fisher only briefly and did not cite Haldane at all, while making it clear that he found Wright's theory the most congenial.[107] His rejection of "essentialism" in favor of "population thinking" was, according to Paul Thompson (2000), associated (perhaps paradoxically) with distaste for the mathematical view of the world. Mayr's own empirical work and general knowledge of the subject had already convinced him that geographic isolation is the key factor in the speciation of birds. Now thanks to Dobzhansky, he had a theory to support that view: Sewall Wright said evolution should be faster in small populations, "and this is exactly what we find"—in the West Indies, Solomon Islands, Galapagos Islands, and Hawaii. Alfred Kinsey found this true for gall wasps, W. F. Reinig for bumblebees.[108] But now Mayr put his own stamp on the idea with a further twist:

> The reduced variability of small populations ... [is due] sometimes to the fact that the entire population was started by a single pair or by a single fertilized female. These "founders" of the population carried with them only a very small proportion of the variability of the parent population.

On his early career and environment, see Mayr (1980a, 1992a, 1999); Hagen (1982); Hapgood (1984); Greene and Ruse (1994); and Provine (2004).

105. Guterman (2005, A17) and Yoon (2005).

106. Mayr and Provine (1980); Mayr (1992); Sloan (1985); Greene (1994); Greene and Ruse (1994); Johnson (2005); and Winsor (2005).

107. Mayr (1942, chapter 1). "The mathematical calculations of Sewall Wright and the theoretical analysis that preceded them have been of particular value" (Mayr 1942, 217). He was strongly influenced by Dobzhansky's publications and lectures (Mayr 1980b). He rejected the selectionist view on polymorphisms of Fisher and Ford (Mayr 1942, 75). Later he criticized "beanbag genetics" as used, for example, by Haldane in his papers on the "cost" of natural selection; see Ewens (1993). Mayr's 1991 hindsight view of drift is quoted next.

108. Mayr (1942, 236); Kinsey (1936, 1937); and Reinig (1939). For a survey of Mayr's views on genetics and speciation, see Provine (2004).

This "founder" principle sometimes explains even the uniformity of rather large populations.[109]

As examples of the founder principle, Mayr cited his own work on the reef heron (*Demigretta sacra*), whose color is always gray on some islands but both gray and white on others.

Mayr's support for isolation as the major cause of speciation in birds did not prevent him from endorsing the importance of natural selection in other cases, while cautioning readers to be skeptical about the adaptive significance of some differences invoked by extreme selectionists.[110]

Mayr took a firm stand in favor of another postulate of the natural selection hypothesis: there is no qualitative difference between macroevolution and microevolution. He wrote:

> All the properties and phenomena of macroevolution and of the origin of the higher categories can be traced back to intraspecific variation, even though the first steps of such processes are usually very minute. (Mayr 1942, 298; see also 291)

Huxley was convinced by Mayr's argument for isolation as the major cause of speciation and praised Mayr's "definitive refutation of Goldschmidt's heterodox views on speciation" (1943, 348).

But Mayr's early support for Wright's general theory was much less firm, according to his recollections half a century later. In a letter to Michael Ruse dated November 20, 1991, Mayr wrote:

> In my 1942 book, in order to be "modern," I quote Sewall Wright copiously. However, in my actual thinking and working I was very much op-

109. Mayr (1942, 236). In another nod to the mathematical approach, he credited Wright for the idea that *fluctuations* in population size can produce the same effect. Mayr later credited B. Rensch with an earlier recognition of the Founder Principle (Mayr 1980d, 26).

110. For caution against extreme selectionism, see Mayr (1942, 75, 96). As good examples of selection he mentions the Palearctic larks (*Alaudidae*), "the first birds in which correlation between soil and coloration was discovered" (by Hartert) (Mayr 1942, 86); Sumner's work on *Peromyscus* (Sumner 1932); and Benson's study of melanistic rodents in lava flows (Benson 1933). Concerning the research of Dice and Blossom, Mayr argues that the color can't be due to climate because the blackest races sometimes occur near the palest, so it "is obvious that selection by predators must have played an important role" (Mayr 1942, 87). Rensch's rules, sometimes rejected because he first used them as evidence for Lamarckism, can be interpreted as a result of natural selection (Mayr 1942, 89). Huxley's "clines" for character gradients correlated with climatic factors are a consequence of natural selection, though in some cases there seems to be no adaptational significance (Mayr 1942, 95).

posed to him. And I fought Dobzhansky all along when he wanted to believe in the neutrality of the human blood group genes and the Drosophila gene arrangements. (Ruse 1996, 414)

During the 1940s, Mayr expressed increasingly strong support for the natural selection hypothesis. In 1945, discussing Epling's data on gene arrangements in *Drosophila* (see chapter 12), he made what he later called a prediction that the evolution of these arrangements would be found to be influenced by natural selection (Mayr 1963, 207). A 1949 paper summarized the "impressive body of evidence" for selection's effects in both zoology and botany; even those like Goldschmidt who denied that it can explain macroevolution admitted its adaptive role in local populations (Mayr 1949b, 519).

Will Provine sees a turning point toward selectionism in a 1950 paper by Mayr and Stresemann, which abandoned the position in *Systematics and the Origin of Species* that "most conspicuous color polymorphisms in birds and other animals were nonadaptive" and that they are "accidents of variation and without selective significance" (Provine 1986, 453). Instead, Mayr and Streseman wrote, "Since recent genetic evidence indicates that alleles involved in a balanced polymorphism have different selective values, it seems probable that many subspecies and species characters that have heretofore been considered as 'neutral' are controlled by genes which differ in their selective values" (Mayr and Streseman 1950, 299).

In his 1963 book, *Animal Species and Evolution*, whose influence on naturalists was attested by Stephen Jay Gould (2002, 535), Mayr included a vigorous attack on the use of the genetic drift concept in evolutionary biology. The term is used in so many contradictory ways that it "has been rather discredited":

> During the period from about 1935 to 1955 it was fashionable to attribute puzzling evolutionary changes to "drift" or to the Sewall Wright effect in the same manner in which the preceding generation of evolutionists had explained similar changes as due to "mutation." (Mayr 1963, 204)

Drift is often postulated in cases of supposed selective neutrality, but such cases are actually quite rare:

> Selective neutrality can be excluded almost automatically wherever polymorphism or character clines are found in natural populations. This clue was used to predict the adaptive significance (previously denied) of the distribution pattern of the gene arrangements in *Drosophila pseudoobscura*

(Mayr 1945) and of the human blood groups (Ford 1945). Virtually every case quoted in the past as caused by genetic drift due to errors of sampling has more recently been interpreted in terms of selection pressures. (Mayr 1963, 207)

Why did Mayr change his position on drift? We have at least two possible answers. According to his published remarks, it was because of the evidence. However, according to his letter to Ruse, he had always rejected drift, so we may conjecture that he went along in 1942 because he did not want to challenge what appeared to be scientific orthodoxy; later, when he had established his reputation, he could afford to do so. Here, as in other cases, we cannot prove why a particular scientist expressed a particular view—at least not without much fuller evidence.

Paradoxically, Mayr (1980d, 26) continued to maintain the validity of his own founder principle, which could be regarded as a special case of genetic drift, perhaps because it did not depend on the assumption of selective neutrality.

10

Simpson: No Straight and Narrow
Path for Paleontology

In his 1944 book, *Tempo and Mode in Evolution*, George Gaylord Simpson presented the evolutionary synthesis as a theory that is compatible with, if not derivable from, the findings of paleontology.[111] In particular, he showed that those findings did not require the invocation of orthogenetic rectilinear trends. This kind of trend "is a product of the tendency of the minds of scientists to move in straight lines [rather] than of the tendency for nature to do so" (Simpson 1944, 166). The only specific evidence for natural selection, admittedly indirect, was the example of the horse. Evidence that paleontologists were abandoning orthogenesis in favor of natural selection may be found in the contributions of D. M. S. Watson and Alfred Romer to the 1947 Conference on Genetics, Paleontology, and Evolution (Jepsen et al. 1949, 49, 58, 107–9). Simpson favored Wright's drift over a purely selectionist explanation, a preference that was reversed in his later book *The Major Features of Evolution* (1953a).[112]

111. Simpson, quoted in Mayr (1980c, 452–63). This article, based on Simpson's answers to questions from Mayr (because Simpson could not attend the workshops or write an original article for it), seems to me especially valuable because it forces Simpson to remember influences and to articulate opinions that he might not have included if left to prepare a paper on his own. Of course that does not make it more accurate or objective.

Simpson considered Schindewolf's 1936 book, along with Goldschmidt's *Material Basis of Evolution* (1940), to be "seriously misleading. . . . Their theoretical views were untenable. The necessity of considering such opposing views was a definite factor in the formulation of the synthetic theory" (Mayr 1980c, 457).

In his introduction to the 1993 translation of Schindewolf's 1950 book, Stephen Jay Gould says that in 1965, his advisor, Norman Newell, called Schindewolf "the world's greatest living paleontologist," despite his opposition to the evolutionary synthesis which Newell himself advocated (Gould 1993).

112. Simpson, quoted in Mayr (1980c, 456). He continues: "My manuscript was completed before Mayr's book appeared, and thereafter I could make no changes because I was overseas in the army until late in 1944, after my own book was published" (457).

According to Stephen Jay Gould and others, Simpson singlehandedly brought paleontology into the evolutionary synthesis with *Tempo and Mode*, followed by a more popular exposition, *The Meaning of Evolution* (1949b), and a series of articles.[113] At the same time, he strengthened the synthesis itself by adding a quantitative *temporal* dimension for evolution, which laboratory and field studies of natural selection could not supply.

But Gould, a paleontologist who had deviated from the evolutionary synthesis by proposing a theory of "punctuated evolution," complained that the synthesis had evolved into a rigid dogma that stressed natural selection to the exclusion of all other alternative or contributory factors. As a paleontologist, Gould was particularly concerned with the role of Simpson, who had argued that fossil evidence, especially for horses, that had been used to support orthogenesis, could be better explained by natural selection. Gould wrote that Simpson's 1953 book, *The Major Features of Evolution*,

> differs from *Tempo and Mode*. . . . It . . . displays some subtle but important shifts in theoretical emphasis and content. These shifts mirror some general trends in the modern synthesis, as its theory won adherents, gained prestige, and (unfortunately in some respects) hardened. In particular, increasingly exclusive reliance on selection-toward-adaptation (for Simpson, in the gradual, phyletic mode), coupled with a greater willingness to reject alternatives more firmly than the evidence warranted, marks both Simpson's new book and the growing confidence of the synthetic theory in general. . . . Though he chides others (quite properly) for assuming that structures are inadaptive because they cannot imagine a use for them, he often constructs adaptive scenarios, in the speculative mode and on the opposite (and equally invalid) assumption that prominent features must have some immediate use. (Gould 1980, 166–7)

Moreover, Simpson had rejected Wright's genetic drift (which he had favored in *Tempo and Mode)*, concluding that (in Gould's words) "it could not trigger any major evolutionary event."[114] For Gould, this was tantamount to devaluing the profession of paleontology: Simpson had

---

113. Laporte (2000). For details of one important effort toward collaboration of paleontologists and systematists, see Cain (2004). But moving paleontology in a new direction was a slow process, as Simpson himself pointed out (1950). According to Bernhard Rensch, "Lamarckian explanations of speciation were widely held [by German paleontologists] in the period 1920–1940 (and sometimes even later)"—even in 1952 (Rensch 1983, 35). On the diversity of responses to *Tempo and Mode*, see Cain (2003).

114. Gould (1980, 168). The statement he quotes from Simpson is less dogmatic: "Genetic drift is certainly not involved in all or in most origins of higher categories, even of very high cate-

unified paleontology with evolutionary theory, but at a high price indeed—at the price of admitting that no fundamental theory can arise from the study of major events and patterns in the history of life. Why be a paleontologist if all fundamental theory must arise elsewhere? (Gould 1980, 170)

Gould, although extremely interested and well read in the history of evolutionary biology, was certainly not an unbiased interpreter of primary sources. The acceptance of his own theory of punctuated equilibrium would require the rejection of extreme selectionism. Reinstating random genetic drift as a significant factor in evolution might help his case, although it would have been risky to face quantitative scrutiny by population geneticists versed in the mathematical theories of drift developed by Fisher and Wright. More to Gould's advantage would be a resurrection of Goldschmidt's macromutations, but as empirical phenomena without Goldschmidt's theoretical speculations (Gould 2002, 451–66). With this in mind let us look at two of Simpson's views *not* quoted by Gould.

In 1967 Simpson stated that the theory of sudden major changes (Schindewolf, Beurlen, and Goldschmidt) was

> conclusively refuted by Rensch. In fact there is now almost no support for that view except by a few philosophers not sufficiently acquainted with scientific data on evolution.

In particular, Goldschmidt's "systemic mutations" have

> never been observed and . . . need not be taken seriously if, as is the case, the phenomena that they were postulated to explain can be explained in terms of known processes and forces. . . . In some of his latest work Goldschmidt implicitly retreated from his position. (Simpson 1967, 231–2)

But at the same time, he was rejecting macromutations and orthogenesis by again allowing a role for genetic drift. Having (to Gould's dismay) hardened his view against drift in the 1950s, he restored it as a secondary factor in the 1960s. Drift "plays a role, perhaps a minor one," although Mayr's founder principle had more influence on evolution (Simpson and Beck 1965, 431).

---

gories such as classes or phyla" (1953a, 355). The next sentence, not quoted by Gould, is: "It is not positively known to have been involved in any instance." Michael Ruse, in his review of Gould's *Structure of Evolutionary Theory*, criticized his "notorious" hardening metaphor—"Gould's facts are made to fit the metaphor, no matter what" (Ruse 2003, 397).

It provided a useful weapon against the orthogenesis and teleology that still contaminated the writings of some of his fellow paleontologists.

In the 1960s Simpson was also rejecting in another way the hardening of the synthesis that Gould protested: "The existence and persistence of apparently random as well as clearly oriented features in evolution [is] an unanswered argument against theories demanding the reality of purpose or the existence of a goal in evolution. It equally renders untenable all the other theories that attempted to explain evolution by the dominant or exclusive action of one single principle or another, such as the [n]eo-Darwinian insistence on natural selection as essentially the whole story. . . . Nonadaptive and random changes have . . . a bearing on changes in broad types of organization, the appearance of new phyla, classes, or other major groups" (Simpson 1967, 230–1).

11

....

Stebbins: Plants Are
Also Selected

The process of bringing the major biological disciplines into the evolutionary synthesis was completed, according to most historians of the subject, by the publication of *Variation and Evolution in Plants* by G. Ledyard Stebbins Jr. in 1950.[115] Research in botany had been responsible for two major obstacles to the acceptance of the modern natural selection hypothesis in the early twentieth century: de Vries's mutation theory, based on his experiments with the evening primrose (*Oenothera*), and Johannsen's pure line theory, based on research with beans.[116] Yet "with the exception of the fact that genes lie on chromosomes... every other significant fact about genetics needed for the synthetic theory was worked out from research on higher plants" (Stebbins 1980, 139). Thus by 1950, a major statement about evolutionary theory from a botanist was long overdue.

Addressing the assertion (see chapter 1) that natural selection is only negative and cannot create anything new, Stebbins invoked a persuasive metaphor: it is negative only in the sense that a sculptor's creation of a statue by removing chips from a block of marble is negative.[117]

---

115. Stebbins (1950) and Smocovitis (1988, 1997, 2006). On the role of hybridization and polyploidy, which are especially important for plants, see Anderson (1952) and Kleinman (1999).

116. "De Vries's theory that new species result from single mutations—a theory based on *Oenothera*—was quite a roadblock for botanists, as was Johannsen's claim that two kinds of variations exist and that all that selection did was to separate pure lines. I used to tell my students that Johannsen was a Dane looking at the Danish landscape with its nice neat gardens full of pure lines of peas and beans. His world included no conception of the confusion that exists in a tropical rain forest or in the waste lots of the subtropics, where weeds are coming in and the entire ecosystem is disturbed. He thought that his world was the natural world" (Stebbins 1980, 146). See also Stebbins (1950, 101–2) and Darlington (1980, 76–79). Stebbins (1980) notes the role of polyploidy in *Oenothera* and other plants but does not admit that it gives any support to the theories of de Vries and Goldschmidt.

117. Stebbins (1950, 102). The sculptor metaphor was also invoked by many other authors; for example, Lerner (1959) and Mettler and Gregg (1969). Holmes (1948b) argued at length

But the direct botanical evidence for natural selection was still meager. Among cross-fertilizing plants, only two experiments demonstrated it: Sylvén showed in 1937 that clover adapts to a colder climate by elimination of the less hardy plants, and Clausen, Keck, and Hiesey (1947) presented data on adaptation of ecotypes of the coast tarweed of California (*Hemizonia augustifolia*), of a sunflower hybrid, of the tidy-tip (*Layia platyglossa*), and of rust (*Potentilla gladulosa*). Stebbins mentioned a few other experiments on homozygous self-fertilized species. Later in the book, he cited research by Harlan and Martini (1938) on barley as "the most extensive experiment yet performed on natural selection in the higher plants," which confirmed the prediction by Fisher and Haldane for the "survival of alleles in a heterozygous population. Selection acts very rapidly at intermediate gene frequencies, but more slowly at very high and very low ones."[118] In the absence of extensive data on plants, Stebbins included Demerec's research on the resistance of the bacterial species *Staphylococcus aureus* to penicillin (see chapter 14), Gershenson's (1945) experiments on seasonal and annual variations in melanic forms of hamsters, Dobzhansky's work on *Drosophila pseudoobscura*, and Dubinin and Tiniakov's results for *Drosophila funebris*. The only comparable evidence for higher plants was that of W. B. Kemp (1937) for mixtures of grass and clover, separated into nearby fields subjected to different conditions (used for grazing or for hay).

While Stebbins had defended the Sewall Wright effect in 1944,[119] by 1950 he had lost his enthusiasm for it. In most cases, he believed, the population is not small enough for it to be significant. It is likely, but not proved, that it applies to some plants in Hawaii (*Gouldea, Cyrtandia, Bidens*), and it is very likely that it affects land snails. "The only plant example known to the writer in which the action of random fixation or drift seems to have taken a prominent part in the differentiation of small isolated populations is that of the complex of *Papaver alpinum* in the Swiss Alps, as described by Fabergé (1943)" (Stebbins 1950, 145–7).

---

that natural selection is not purely negative since it includes variation. The assertion that natural selection is only a negative factor, so it cannot explain the adaptation of the organism that does evolve, is no longer taken seriously by biologists since there is good evidence of ample variation in natural populations and enough favorable new mutations to allow progress by selection. But the issue is still debated by philosophers (Nanay 2005).

118. Stebbins (1950, 109) and Harlan and Martini (1938). This is a rather mystifying paper: not only do the authors fail to cite Fisher and Haldane, but they also give no explanation of where they got the "theoretical curve" to which they compare their data. Stebbins had also cited Harlan and Martini, along with Sylvén, Clausen et al. (1947), and his own research on the competition between diploid species of grass and the autotetraploid derived from it, in Stebbins (1949).

119. Stebbins (1944); see Cain (2004, 68–69).

12

....

Chromosome Inversions
in *Drosophila*

During the 1940s and 1950s, several biologists obtained empirical results that seemed to show that certain characters, previously considered nonadaptive and therefore attributable to the effects of random genetic drift, were instead primarily controlled by selection. First and perhaps most important in its impact on the views of evolutionists was the series of observations conducted by Dobzhansky and his colleagues (including Carl Epling and Sewall Wright) on the genetics of wild populations of *Drosophila pseudoobscura*. The background and motivation for this research are described in detail by Provine (1986); Provine shows that Dobzhansky originally wanted to find a definitive empirical confirmation of Wright's theory of genetic drift but ended up doing just the opposite.

Epling's (1944) detailed observations on the third chromosome of *D. pseudoobscura* suggested that different gene arrangements had evolved from a single ancestor by inversion of chromosome Chapters. His results were discussed in a 1945 symposium by Ernst Mayr, G. G. Simpson, and G. L. Stebbins.[120] Epling had assumed that the gene arrangements were equal in selective value, but Mayr argued that "each gene arrangement . . . may have different selective values at each locality. This is a consequence of the fact that the reduction of crossing over in the inverted Chapters prevents the free recombination of genes" (Mayr 1945, 74). Stebbins agreed that some gene combinations would have a

> selective advantage under certain ecological conditions, an assumption
> for which Dobzhansky . . . has obtained some indirect evidence. I agree

120. For further details, see the correspondence published in *Bulletin* no. 4 (November 12, 1944) of the Committee on Common Problems of Genetics, Paleontology, and Systematics, reprinted by Cain (2004, 87–107).

....

with Mayr that ... the maintenance of any distribution pattern over periods of thousands or millions of years would be improbable unless the dispersal of the various chromosomal types were restricted by the selective activity of the environment.... Epling ... now holds the same belief. (Stebbins 1945)

Dobzhansky himself had already come to the same conclusion, as he explained later (Dobzhansky 1951, 1971). The chromosome inversions in *D. pseudoobscura* are examples of *adaptive polymorphism*, which can be observed in field studies but also produced under laboratory conditions where the relevant variables can be controlled:

> Two species, A and B, can be sympatric [occupy the same territory] only provided that the environment in a territory which they inhabit is heterogeneous. The heterogeneity may be spatial or it may be temporal.... A may be better adapted than B in summer, while B is superior to A during the winter season. (Dobzhansky 1951, 109)

By shifting the balance back and forth between A and B in response to seasonal changes in the environment, natural selection allows the organism to exploit the available resources more efficiently than if only one of the two genotypes, A or B, were allowed to survive.

These observations made use of the giant chromosomes in the salivary glands, which have a visible pattern reflecting the gene arrangement in the chromosomes. Usually the third chromosome was observed. There are various possible rearrangements, or inversions, of the chromosome segments, producing a polymorphism. For example, the sequence ABCDEFGHI may be broken between A and B and between E and F, giving A BCDE FGHI. Then the middle sequence BCDE is inverted, giving AEDCB FGHI. Finally the sequence DCBFGH is inverted, leading to AEHGFBCDI.

> The first can arise from the second or give rise to the second through a single inversion. The same is true for the second and the third. But the third can arise from the first, or vice versa, only through the second arrangement as the probable intermediate step in the line of descent. If we find in natural populations of some species only the first and the third arrangements, it is probable that the second remains to be discovered, or at least that it existed in the past.... The existence of previously unknown gene arrangements in *Drosophila pseudoobscura* and *D. azteca* was predicted with the aid of the theory of overlapping inversions, and most of these pre-

dictions were subsequently verified by discovery of the requisite inversions in nature. (Dobzhansky and Sturtevant 1938; Dobzhansky 1941a).[121]

Note that we have here an example of a confirmed novel prediction, based on the hypothesis that the chromosomes have evolved but not necessarily by natural selection (Dobzhansky and Sturtevant 1938, 33). It is comparable to the prediction of "missing links" from the general hypothesis that later organisms have evolved (by some process) from earlier ones, repeatedly confirmed in the late nineteenth and twentieth centuries by discoveries of fossils.

Dobzhansky recalls, "It seemed at first that the chromosomal polymorphism had no adaptive significance. It is now known that the contrary is the case" (1951, 114). The importance of this realization is reflected in the reorganization of the third edition (1951) of *Genetics and the Origin of Species*: the inversion experiments are described in a chapter now titled "Adaptive Polymorphisms" (instead of the noncommittal "Chromosomal Changes" in the second edition), which is now placed immediately after the chapter on "Selection," rather than before it.

Since David Wÿss Rudge (2000a) has given a detailed account of these crucial observations and experiments, I need present only a brief summary here. Experiments with population cages showed a seasonal variation of the frequencies of the different patterns. Starting with an experimental population containing 11 percent of "Standard" (ST) and 89 percent of Chiricahua (CH) chromosome, within four months the frequency of ST rose to 70 percent.

> It is clear, that the carriers of ST chromosomes had some adaptive advantages under the conditions of the experiment over the carriers of CH chromosomes, and that the rapid increase of the incidence of the former was caused by natural selection. But if so, why has selection failed to eliminate CH chromosomes altogether? The establishment of the equilibrium, at which both ST and CH chromosomes are present in the gene pool with definite frequencies, is due to the chromosomal polymorphism being balanced, because the heterozygotes (individuals having one ST and one CH third chromosome) are superior in adaptive value to both chromosomal homozygotes (ST/ST and CH/CH). (Dobzhansky 1951, 114–5)

---

121. The quote including the two citations at the end is from Dobzhansky (1951, 110–2); his figure 3 (111) shows the proposed mechanism for the inversions. For description of the origin and results of the Genetics of Natural Populations series, see Provine (1981) and Lewontin (1981).

What does this have to do with evolution? "Since the gene arrangement in a chromosome is a hereditary trait, we are dealing here with genetic changes in the constitution of a population. These are evolutionary changes by definition. Furthermore, these changes are brought about by natural selection." Moreover, it is an evolutionary process that can be quantitatively studied under controlled conditions: "The speed of the changes observed in nature during the hot season can be accounted for if the chromosomal types have adaptive values which they are observed to have in experimental laboratory populations at 25°C" (Dobzhansky 1951, 118).

According to Provine, the first public indication of Dobzhansky's shift toward the NSH was in 1943, where he

> breaks rather abruptly with the past: because of cyclic (seasonal) and year-to-year changes in the frequencies of ... gene arrangements ... he concludes that these gene arrangements affect the adaptive values of their carriers. Cyclic changes ruled out chance, and mass migration was ruled out by the studies on dispersal. Thus the proportions of these gene arrangements in a population were said to be subject to a surprisingly intense natural selection. (Provine, in Dobzhansky 1981, 303)

But Provine notes Dobzhansky "had concluded that the inversions were under selection two years before the appearance" of this paper, as is shown by his letter to Sewall Wright dated May 4, 1941.[122] As sometimes happens in a fast-moving research project, the public record of a scientist's position does not always keep up with changing views expressed in private. For our purposes, the public position as well as the private view is important, since for a scientist as influential as Dobzhansky, the public position is likely to persuade others.

In a 1946 paper, Dobzhansky asserted on the basis of his *Drosophila* experiments that natural selection could cross the supposed barrier between micro- and macroevolution:

> Some of the chromosomes obtained by crossing over between the three ancestral wild chromosomes have properties very different from the latter. It is, therefore, possible to "select" products of recombination of the gene complexes that deviate greatly from the ancestral types, being completely outside the limits of variability of these ancestors.

---

122. Provine (1986, 389ff). The discussion toward the end of the paper by Dobzhansky and Sturtevant (1938, 61) seems with hindsight to be an intermediate stage in the transition from nonadaptive to adaptive description of chromosome inversions.

This allows "great advances in rebuilding the organism in directions favored by artificial or natural selection" (Dobzhansky 1946, 288). In a review published as the first article in the new journal *Evolution* (1947), Dobzhansky claimed that his own work, along with that of Timofeeff-Ressovsky—who first discovered cyclic seasonal changes in 1940—and the more recent publications of Dubinin and Tiniakov, disproved the previous view that "adaptive evolution in nature is too slow a process to be observed within a human lifetime."[123]

A brief note on these precursors acknowledged by Dobzhansky: N. W. Timofeeff-Ressovsky was a prominent Russian geneticist best known for his research on the reversibility of mutations.[124] In his study of populations of the lady-beetle, he found a polymorphism maintained by climate-dependent selection pressure that favored the red form in winter and the black in summer (Timofeeff-Ressovsky 1940). N. P. Dubinin and G. G. Tiniakov (1945, 1946a, 1946b, 1946c) studied populations of *Drosophila funebris* and found "evidence of an energetic selection which has changed the genetic structure of populations in response to changes in habitats, contrary to the prevailing view that evolution proceeds very slowly" (1945, 572). All these results clearly pointed to natural selection as a powerful process that could produce observable effects in just a few months, with no help from random genetic drift. (Recall that Dubinin himself had proposed a form of drift in 1931.)

According to Dobzhansky's testimony, it was the surprising result of his own experiments that led him to change his opinion about the relative importance of natural selection and genetic drift: in these experiments, drift clearly had a negligible effect, contrary to what was previously believed.[125] On the other hand, John Beatty suggests that Dobzhansky did not have

123. Dobzhansky (1947, 1). While Wright was a coauthor of some of Dobzhansky's papers in the "GNP" series, Provine suggests that they did not completely agree on the interpretation of the results. See the editorial note in Dobzhansky (1981, 470). Burian (1994) suggests that Dobzhansky's rejection of macromutation as a phenomenon not explicable by natural selection should be seen in the context of his relations with his mentor Iurii Filipchenko, who coined the term "macromutation."

124. On the career of Timofeeff-Ressovsky, see Glass (1990); Paul and Krimbas (1992); Junker (1998); and Reindel (2001).

125. Here is Dobzhansky's later account of why he changed his views on natural selection and drift: "In 1939, there was initiated repeated sampling at approximately monthly intervals during the breeding season, of populations of Drosophila pseudoobscura in three ecologically rather different localities on Mount San Jacinto, in California. The purpose of this work was a study of the frequencies of allelism of recessive lethals in the third chromosomes. . . . The results obtained seemed startling: not only were the populations of the three localities, about 15 miles apart, clearly different in the chromosome frequencies, but in two of the localities the chromosome frequencies were changing significantly from month to month. That these changes could

. . . .

enough empirical evidence to justify his change from drift to selection, hence one must invoke Dobzhansky's "values": "He wanted [the world, and in particular the human species] to be full of variation" on which natural selection could act. "Human variation might play a part in enabling the species to respond adaptively to environmental upheavals of various sorts." This would help him make a stronger case against H. J. Muller's "classical" position in the debate about the "genetic load" of mutations in the human population and its significance, and it would allow him to judge "the future of the human species to be quite rosy."[126] Vassiliki Betty Smocovitis supports this interpretation, arguing that for Dobzhansky, "evolutionary models favoring random genetic drift, which enforced a stochastic view of evolution—and culture—would not be favored in a postwar frame of mind seeking to improve the world" (Smocovitis 1996, 131). Gould (2002, 543) thought this was plausible but also pointed out that random drift might be important in a small population of evolutionary biologists: "A reassessment by a few key people ... might trigger a general response." (Dobzhansky, Simpson, and Mayr were all in New York City at that time.)

In the absence of documentary evidence to support the Beatty-Smocovitis interpretation, I prefer to accept Dobzhansky's own statement that the empirical evidence justifies the choice of the selectionist over the drift hypothesis.

---

be caused by natural selection seemed hard to believe.... The possibility that the changes may have resulted from random genetic drift appeared more plausible. A coup de grâce to this surmise was administered by the finding that at least some of the changes are regularly cyclic, following the succession of the year's seasons.... But even this would not have made the selectional explanation believable to many biologists, had it not been possible to reproduce some of the changes in artificial populations in laboratory experiments (Wright and Dobzhansky 1946).... The ... evidence ... clearly shows that many, though possibly not all, natural populations of D. pseudoobscura undergo genetic changes rapid enough to be recorded ... in successive months. It is certain that these changes are brought about by natural selection; random processes may contribute at most minor local fluctuations. The nature of the selective forces, the results of which are so plainly visible, is still conjectural—despite more than thirty years of study" (Dobzhansky 1971, 110–1, 127–8).

For the extension of Dobzhansky's research to Mexican populations of D. *pseudoobscura*, see the paper by Levine et al. in Levine (1995, 120–39).

126. Beatty (1987a, 300, 303, 305)

13

....

Ford: Unlucky Blood Groups

The leader of the Oxford School of Ecological Genetics, E. B. Ford, conducted and sponsored many experiments, testing the roles of natural selection and genetic drift and developing the concept of balanced polymorphism.[127] By 1940 he could claim that this research confirmed the importance of natural selection. Here I discuss just one of his contributions: the prediction, based on Fisher's theory, that human blood groups would be found to be correlated with certain diseases.

In 1940 Ford pointed out (as a consequence of Fisher's "fundamental theorem") that variations in the fitness of one member of a polymorphism caused by changes in the environment should be correlated with changes in its variability. This can be observed, for example, in the butterfly *Papilio dardanus*: fitness depends on mimicry of another species.[128] Conversely, if we find changes in the distribution of members of a polymorphism, we may infer that fitness is due to some as yet unknown environmental factor. Ford proposed to apply this idea to the human blood group polymorphism by predicting a connection between the frequencies of different groups and susceptibility to specific diseases.

There had already been hundreds of reports claiming to find statistical correlations between the ABO groups and various medical conditions or psychological characteristics (including criminality), going back to 1921. But these reports were not considered reliable by modern (i.e., 1940s) standards, because the investigators did not realize that very large samples must be used to obtain significant correlations.[129] Within the community of ex-

---

127. Turner (1987); Ford (1957b; 1980, 341–2); and Ruse (1996a, 457). In a balanced polymorphism—for example, between melanic and nonmelanic moths—the selection pressures are often equally strong on both sides, so it is not obvious how strong they are until changing environment reduces the pressure against one form and allows it to predominate. See the quote in following text. For additional references to experiments whose results disproved the Sewall Wright effect, see Skipper (2000). For further information on Ford, see B. C. Clarke (1995) and Kimler (1983a).

128. Ford (1940a, 505); see also Ford (1937).

129. Hardin (1952, 418) and Garratty (1994, 1996).

....

perts on population genetics, no such correlations were believed to have been demonstrated or likely to be established.[130] Ford made his prediction more explicit in 1945:

> Individuals belonging to the different blood groups are not equally viable, and we may expect elimination to fall upon the AB class.... A valuable line of enquiry which does not yet seem to have been pursued in any detail would be to study the blood group distributions in patients suffering from a wide variety of diseases. It is possible that in some conditions infectious or otherwise, they would depart from their normal frequencies, indicating that persons of a particular blood group are unduly susceptible to the disease in question. (Ford 1945, 85)

Ford's prediction was confirmed in the 1950s.[131] The initial report by London medical researchers Aird, Bentall, and Roberts (1953) stated that persons with blood group A are significantly more likely than those with O to have cancer of the stomach. They did not mention Ford's prediction. In a more comprehensive article the next year, which also reported an association with peptic ulceration, these authors, together with Mehigan, did credit Fisher and Ford with a prediction but did not provide a precise citation (Aird et al. 1954, 318). A team at the State University of Florida published a survey of early results, noting that Ford (1945) "hinted at" a relationship between blood groups and diseases "in suggesting the possible relationship of blood groups to natural selection."[132]

J. A. Fraser Roberts concluded that firm correlations had been established between duodenal ulceration and group O and between pernicious anemia, as well as stomach cancer, and group A; he also found "fairly strong evidence for an association between group A and diabetes mellitus." In a very brief theoretical comment, he did not mention Ford but asserted: "Fol-

---

130. Aird, Bentall, Roberts (1953), and Kitchin et al. (1959). Of course the prediction was based on Fisher's theory. According to Michael Ruse (1996a), such successful predictions helped Ford to get grants to support further research. On Fisher's interest and participation in research on blood groups, see Box (1978, chapter 13).

131. According to Ford (1957b), the first confirmation of his prediction was by Struthers (1951), who studied bronchopneumonia in babies and found the incidence higher with group A than group O. A. E. Mourant (1959), an expert on blood groups, acknowledged the success of Ford's prediction and admitted that "it is unlikely that any genes are completely unaffected by natural selection"(Mourant 1959, 60); yet he insisted that we still don't know "the mechanism whereby blood groups affect liability to the diseases mentioned, if indeed the relation is one of cause and effect" (59). Apparently he would have traded a successful prediction for a plausible explanation.

132. Buckwalter et al. (1956b, 1215); see also Buckwalter et al. (1956a, 1956).

lowing Fisher [1930a] a neutral gene is almost unthinkable. . . . A long continued polymorphism must be dynamic, not static. It must depend on an interplay of selective advantages and disadvantages" (Roberts 1957, 124).

A multiauthor 1989 book on the history of probability portrays the discovery of the blood-group and disease connection, along with Dobzhansky's "rather dramatic turnaround" (triggered by his results on chromosome inversions) and the work of Cain and Sheppard on *Cepaea*, as part of a turning of the tide against drift and "overwhelming support" for selectionism.[133]

Ford was of course entitled to claim victory for his own theory in this particular case (Ford 1957b); at the Darwin centennial celebrations in 1959, he told the audience that his 1945 prediction about the selective control of human blood groups had now been confirmed (Ford 1960, 194). His students and colleagues were also quick to point out the new support for selectionism in the ongoing controversy with Sewall Wright. But Wright had already changed his views by 1951, when he suggested that the distribution of blood group frequencies may be the result of "intergroup selection . . . a certain balance between local isolation and cross breeding" (Wright 1951, 454).

Nevertheless, I could not find any evidence that other biologists outside Ford's group were persuaded to accept his selectionist views *because* of this confirmation, nor that it made any difference whether it was a *novel* prediction rather than a theoretical deduction of known facts.[134]

Before 1950 Dobzhansky had supported Wright's view that drift plays an important role in evolution as an adjunct, though not a substitute, for natural selection (Beatty 1987a), particularly in the evolution of human blood groups (Dobzhansky and Montagu 1947). In his book *Evolution, Genetics, and Man*, Dobzhansky still wanted to explain the differences in blood groups among different populations as a result of genetic drift, ignoring the recent discoveries concerning their connection with diseases, even though he was aware of Allison's work on the connection between sickle cell and malaria (Dobzhansky 1955, 130, 143). In a survey article in *Science*, he stated, "The functional significance of the blood groups in man is still full of uncertainty" (Dobzhansky 1958, 1093), and in another survey article he still found the evidence "inconclusive."[135] In his book *Mankind Evolving*, he cited the evidence for blood group connections with diseases (now including plague, syphilis, and smallpox) but tried to explain them by genetic drift (Dobzhansky 1962, 280).

133. Gigerenzer et al. (1989, 157); John Beatty is probably responsible for this chapter.
134. Sheppard (1958, 1959a); Dowdeswell (1958); and C. A. Clarke (1961, 1971).
135. Dobzhansky (1961); reprinted in Bajema (1971; see especially 12).

In *Genetics of the Evolutionary Process* (the fourth edition of *Genetics and the Origin of Species*), Dobzhansky told the story this way: "The frequent mention of blood groups as examples of adaptively neutral traits goaded many investigators to discover physiological correlates of the blood groups that are not neutral. . . . A relationship among the classical A, B, and AB blood groups and duodenal and gastric ulcers is probably the most thoroughly established." But he did not give Ford any credit for his prediction and downplayed the significance of the discovery that these characters are not adaptively neutral: "The associations thus far discovered are with diseases affecting mainly people of postreproductive ages, so that the Darwinian fitness is but little changed." Moreover, he insisted that the connection with disease could not explain the observed differences in blood group frequencies among local populations, although he did omit the suggestion, made in *Mankind Evolving*, that those differences are a result of genetic drift.[136]

By 1970 Dobzhansky had revived his affection for drift, especially Mayr's founder principle, which he now said was supported by the work of his own group.[137] He denounced the "hyperselectionism" of other evolutionists, although he admitted that the new interest in neutral or non-Darwinian evolution was perhaps just another "swing of the pendulum" (Dobzhansky 1970, 262).

One might suspect that hyperselectionism was just a nationalistic symptom of British distaste for the American drift theory (Gould 1983; Turner 1987). But one of the hyperselectionists identified by Dobzhansky was Ernst Mayr, whose views were discussed in chapter 9. While Mayr did assert that because selective neutrality is rare, "it appears probable that random fixation is of negligible evolutionary importance,"[138] his continued support for his own founder principle would seem to disqualify him as a hyperselectionist.[139] Dobzhansky, in quoting Mayr, had left out part of Mayr's statement, a sentence that explained why he rejected genetic drift as a major factor in evolution: "Virtually every case quoted in the past as caused by genetic drift due to errors of sampling has more recently been interpreted in terms of selection pressures" (Mayr 1963, 207). It is hard to believe that this is literally true, since so many authors had invoked drift to explain so many different examples of supposedly nonadaptive characters. But it does illustrate the rhetorical strategy of the selectionists, followed also by Cain

136. Dobzhansky (1970, 296–8; 1962, 280–3).
137. Dobzhansky and Pavlovsky (1957); Dobzhansky and Spassky (1962); and see chapter 17 herein.
138. Mayr (1963, 211), as quoted by Dobzhansky (1970, 262).
139. Mayr (1963, 204); cf. Mayr (1970, 120).

. . . .

(1951a) and Sheppard (1967): if several phenomena previously ascribed to drift have been shown to be really due to selection (and by implication, the converse has not happened), then it's reasonable to suspect that all other phenomena ascribed to drift are also really due to selection. Of course the rhetorical force of the statement was enhanced if some of these reversals were confirmed novel predictions and if, as Provine argued, "the three most frequently cited examples" of drift—*Drosophila* chromosome inversions, banding and coiling patterns in snails, and blood groups—"were all found to be subject to substantial or even enormous selection pressures" (Provine 1983, 65). (I return to this issue in the final chapter.)

Wright and other theorists who discussed genetic drift often insisted that the fate of such a very small subpopulation was more likely to be complete extinction of the drifting gene: once a gene frequency fluctuates to zero, it can't fluctuate back to a finite value. Blood group researchers found that some subpopulations of American Indians—sixty-two out of 109 surveys, according to Neel and Salzano (1964)—completely lacked the B group. This is probably why, according to Slatis (1964), the distribution of frequencies in this population is often attributed to drift, even though, as Brues (1954) had argued, the fact that four-fifths of all possible combinations of frequencies were missing might better be explained by natural selection.

According to Gigerenzer et al., "While Dobzhansky's selectionist account of chromosome shapes in *Drosophila* has pretty well stood the test of time, the selectionist accounts . . . of blood groups in humans have not. . . . [The Italian biologist Cavalli-Sforza] has successfully accounted for differences in blood group frequencies among different populations in North Italy in terms of drift alone."[140] According to Linda Stone and Paul F. Lurgin, Cavalli-Sforza's first conclusive evidence was published in 1964.[141] Turner (1987) argues that this research provided the missing evidence for drift, but from my viewpoint, it came too late (like Kettlewell's work) to play a decisive role in the debate of the 1940s and 1950s.

The association between blood groups and disease is undoubtedly real (Garratty 2000), but biologists in the 1960s were not able to give a satisfactory explanation for the association. It was pointed out by several experts on population genetics that theoretically, the blood group polymorphism should not be stable, because the heterozygotes do not have a selective ad-

140. Gigerenzer et al. (1989, 157), citing Cavalli-Sforza (1969). On the authorship of this quotation, see note 133.
141. Stone and Luquin (2005, 59–74, 209) and Cavalli-Sforza and Edwards (1964).

vantage over the homozygotes—unlike in the sickle cell and malaria polymorphism, where they do.[142]

Something similar happened with three of the best known predictions of elementary particles in physics: the positron, the meson, and the Ω^- (omega minus meson). In each case, the theory that predicted the particle was soon found to be inadequate and was replaced by another theory (Brush 1993b). Nevertheless, until that happened, the success of the prediction could still be considered evidence for the theory.

142. Allison (1955, 1968); Lewontin (1967; 1974, 235); Mettler and Gregg (1969); and C. A. Clarke (1977, 487).

....

14

....

Resistance to Antibiotics

Another application of natural selection to medicine appeared around the same time as Ford's prediction. René J. Dubos, who was one of the first to discover an antibiotic drug (gramicidin), warned that the microbes attacked by such drugs would develop resistance to them.[143] This phenomenon had already been observed with other drugs and was known as "training" or "fastness," but its cause was not understood. It might have been considered analogous to the well-known phenomenon of resistance to insecticides, although there are significant biological differences between the two (Simon 2003).

In a comprehensive treatise on *The Bacterial Cell*, Dubos reviewed experiments made "to establish whether the resistant bacteria always occur in small numbers during normal growth in the presence of the drug, or whether they are produced only as a response to the presence in the medium of the substance with reference to which resistance develops" (Dubos 1945, 322). In other words, is resistance a Darwinian or a Lamarckian phenomenon?

According to historian Angela Creager, before the 1940s bacteriologists favored the Lamarckian explanation. But the notorious Lysenko affair in the Soviet Union cast suspicion on Lamarckian theories and may have made some scientists more receptive to selectionist theories. Creager considers the 1943 experiment by Salvador Luria and Max Delbrück to be "one of the first clear demonstrations that inheritance in bacteria was not Lamarckian" (Creager 2007, 166).

Dubos (1945) concluded that the evidence favors the Darwinian explanation of resistance: it is due to selection acting on the preexisting variability of the microbes. Those that happened to be resistant to the drug could survive and pass on this attribute to their descendants.

By 1944 it was "a well established fact that strains of bacteria resistant to various sulfa drugs, as well as strains resistant to penicillin, may readily be obtained by growing bacteria in media containing increasingly higher con-

---

143. Dubos (1942, 1944) and Moberg (1999).

....

centrations of the respective chemicals." What was not established, despite the Luria-Delbrück experiment, was whether the resistance was induced by the action of the antibiotic or was already present in some of the bacteria as a result of earlier mutations. This question was answered in a classic experiment in 1944 by Milislav Demerec at Cold Spring Harbor. Demerec, using strains of *Staphylococcus aureus*, found evidence that "makes it probable that the second [Darwinian] alternative is correct": resistance of bacteria to antibiotics is a result of selection.[144]

Dobzhansky published, in *Scientific American*, a good popular explanation of Demerec's experiment as a prominent example of "evolution in the laboratory." For those readers who were following a little too recklessly the neo-Darwinian command "Make love, not war," he noted that "in certain cities penicillin-resistant gonorrhea has become more frequent" (Dobzhansky 1950b, 33, 35).

Dobzhansky supported the Darwinian view over the Lamarckian view in the third edition of his influential book, *Genetics and the Origin of Species* (1951), which "made antibiotic resistance a crucial—and observable—example of natural selection at work" (Creager 2007, 176).

Further confirmation came from a classic experiment by Joshua L. Lederberg and Esther M. Lederberg: They showed that bacteria could display an inherited resistance to penicillin, even if they had never been directly disposed to it. The mutations were not *caused* by the penicillin but were already present, thus excluding a Lamarckian interpretation.[145] Although they don't say they are confirming a prediction by Dubos, Joshua Lederberg studied with Dubos at Columbia and later recalled that Dubos's *Bacterial Cell* "is the work from which I can say I learned most of the microbiology I know.... This work was the launching pad for my own investigations" (Lederberg 1990, introduction).

Although bacterial resistance to antibiotics cannot be called a novel prediction from natural selection, its confirmation did provide important support for evolutionary theory, especially for those scientists and physicians who were interested in practical applications of the theory. Dubos himself seemed to regard it not so much as an important scientific discovery but more as an illustration of his general philosophy: humans should not try to conquer nature (in this case by using powerful drugs) but should try to live in harmony with it.[146]

---

144. Demerec (1945, 16, 23); see also Creager (2007, 167ff).

145. Lederberg and Lederberg (1952) and Lederberg (1989); for a textbook description of the experiment, see Curtis (1968, 695–6).

146. Cooper (1998) and Moberg (2005).

15

....

Two Great Debates:
Snails and Tiger Moths

One of the chapters in Julian Huxley's book *The New Systematics* (1940) was a report by Cyril Diver on his studies of two closely related species of snails, *Cepaea nemoralis* and *Cepaea hortensis*, living in the same areas in England and elsewhere in the United States and Europe. He attributed the split of the presumed earlier single species to "random differentiation in small partially isolated populations"—as described by Wright's theory, presented in the same book—and *not* to pure geographical isolation or natural selection (Diver 1940, 327).

After the end of World War II, Arthur J. Cain and Philip M. Sheppard at Oxford collaborated in a new investigation of *Cepaea nemoralis*, designed to test Diver's conclusion and perhaps to settle the Ford versus Wright debate of the late 1940s about the relative importance of natural selection and random drift. This was the beginning of what Provine, in his biography of Wright, calls the "Great Snail Debate." Since he has described it in great detail, I need only summarize his account and conclusions.[147]

Cain and Sheppard, in their first paper on *C. nemoralis*, found (contrary to Diver) that color and banding patterns "have definite selective value" in connection with predation by thrushes. They concluded that speciation should be ascribed to natural selection, not to genetic drift, and that other cases ascribed to genetic drift should be reexamined (Cain and Sheppard 1950, 275). According to Provine, this result did not play the role in the selectionist versus drift debate that might have been expected. Since Dobzhansky and Mayr had used Diver's results to support their pro-drift (non-adaptationist) views in the early 1940s, Fisher, Ford, and other selectionists would have anticipated that this evidence against the importance of drift would be considered evidence against Wright's theory of evolution and that Wright would therefore try to defend his theory by arguing that gene fre-

147. Provine 1986, chapter 12); see also the obituary of Sheppard by C. A. Clarke (1977).

....

quencies were strongly influenced by drift. But in fact, "even before Wright read the Cain and Sheppard [1950] paper on polymorphism in *Cepaea*, he was prepared to believe that the gene frequencies ... were governed primarily by natural selection rather than by random genetic drift" (in the case of "conspicuous" polymorphisms, according to Provine), as shown by a letter from Wright to Cain in November 1950. Even after Maxime Lamotte published his own results on *Cepaea* populations in France, which he interpreted as supporting Wright's theory of genetic drift, Wright insisted that while genetic drift is a real phenomenon, it is not important for evolution, except as it may interact with natural selection; he was not really interested in invoking drift to explain neutral characters, which other biologists seemed to think provided its most useful application.[148]

Cain also started a mini-debate in *Nature* when he criticized biologists such as G. S. Carter, who postulate "drift" just because they "personally cannot see" any possible selective value in a variation. Then using the rhetorical strategy later employed by Mayr (see chapter 13), Cain asserted, "So far, every supposed example of random variation [or "genetic drift"] that has been properly studied has been shown to be nonrandom."[149] Carter, who had invoked genetic drift in his recent book on animal evolution (1951a), tried to turn the tables by accusing Cain of claiming that "selective value should be assumed in all characters until the contrary is proved" (Carter 1951a). Cain denied taking that position, maintaining instead that one should not assume either randomness or selection without proof. He did not claim that there is *no* evolution of nonadaptive characters (Cain 1951b). Carter replied, "I am glad that Cain admits that genetic drift may be a real factor in evolution.... We should ... regard it and selection as equally possible explanations when neither is proved" (Carter 1951c). In the second edition of his book, Carter retreated a little by adding the statement "Adaptation of the demes to small differences in their environments may often be more important than drift" and by adding a reference to the 1950 Cain and Sheppard paper.[150] In a 1957 book on evolution, Carter omitted genetic drift entirely.

According to Sir Cyril A. Clarke, the American opposition to the nat-

148. Provine (1986, 441–9 [and references cited therein]; quotation on 441); Lamotte (1951, 1952, 1959); see also Lamotte (1950), a less well-known paper in which he discussed selection, and his discussion remark published at the end of Sheppard (1955). According to Millstein (2007a), Lamotte, unlike most other researchers at the time, "was able to make a good case for a significant role for drift, by studying many populations of different sizes; he showed that the fluctuations were greater in smaller populations, as predicted by theory." Millstein (2007b) gives a detailed account of the work of Cain and Sheppard, as well as of Lamotte.

149. Cain (1951a, 424). Turner calls this the "fairies in the garden" argument against drift (1987, 342).

150. Carter (1954, 247, 262).

ural selection hypothesis, in particular the belief that polymorphisms are adaptively neutral, was still strong in 1959. Sheppard's paper, presented at the Cold Spring Harbor Symposium in that year, which was "the only contribution which gave facts and reasons against neutrality, was deleted by the editor [of the published proceedings]. This injudicious act roused Philip to fury" (Clarke 1977, 477).

Although the Great Snail Debate continued through the next few decades, it was considered in the 1960s to have produced a victory for natural selection over genetic drift. As Stephen Jay Gould wrote in 1982, it was "one of the centerpieces of the adaptationist program"; twenty years later, he still recalled it as a major victory for adaptationism (Gould 2002, 541).

....

Fisher and Ford conducted another test of Wright's genetic drift theory using two colonies of the moth *Panaxia dominula* (scarlet tiger moth), conveniently located near Oxford. They measured fluctuations in the frequency of a heterozygous form, *medionigra*, easily recognizable by the colored wing pattern it produces. They reported in 1947 that these fluctuations were much larger than estimated from random drift theory and hence must be caused by changes in the selective forces. Wright, in his reply (1948), complained that his theory had been misrepresented and that furthermore, it was not legitimate to assume that any effect not due to drift must be due to selection. There were further replies back and forth, leading not to a resolution of the controversy but to great animosity among its participants and to some stimulus to useful further research.[151]

151. Provine (1986) gives a detailed account of this controversy through the 1950s, and Skipper (2000) follows it through the 1990s.

....

16

....

Dunkers and Drift

The result of Fisher and Ford on *Panaxia dominula*, showing that fluctuations in the gene frequency of the medionigra gene are "too large
for such shifts to be attributable to genetic drift [and therefore are] due to
selection[,] . . . has no direct bearing on the question whether the fluctuations at some other locus are attributable to one or the other," according to
a 1952 paper by Bentley Glass, Milton S. Sacks, Elsa F. Jahn and Charles
Hess at the Baltimore Rh Typing Laboratory of the Johns Hopkins University and the University of Maryland. "The real difficulty, as Sheppard
[1951] has pointed out, is to be sure that characters of neutral survival value
actually are such, especially since survival value undoubtedly . . . itself varies
from year to year and place to place." It might be more useful to investigate
human populations known to be derived from very small breeding groups;
with an effective size of 200 or fewer members, "an allele would require a
selective advantage of 0.01 to avoid the actions of drift, and consequently
many genes might be effectively neutral." (Glass et al. 1952, 145, 146)

Such a group is the "religious isolate" known as the Dunkers, living in
Franklin County, Pennsylvania. They descended from a sect of the German
Baptist Brethren, which migrated to America in the early eighteenth century and then split up into three groups in 1881. Glass and his colleagues
chose to measure traits believed to be nonadaptive, including the ABO
blood groups. They found variations in gene frequencies, as compared with
typical samples of populations in Germany and the United States: 59 percent had A blood group in the Dunkers, significantly greater than the 45
percent in Germany and 40 percent in the United States, and Dunkers had
correspondingly smaller frequencies in B, AB, and O. They concluded that
these and other variations "are most reasonably attributable to genetic drift"
(Glass et al. 1952, 158). In a 1954 review, which included research on other
isolated human groups, Glass stated: "It is hard to see how any factor except
genetic drift could possibly account for such divergent gene frequencies as
those of the Eskimos of Thule in Northern Greenland. . . . The Polar Es-

....

kimos have a very high frequency of the ABO blood group allele I^0 and a correspondingly low frequency of the allele I^A" (Glass 1954, 130).

Neither paper from the Glass group cited Ford's 1945 prediction that the human blood groups would be found to have selective advantages or disadvantages or the 1953 confirmation of that prediction by Aird, Bentall, and Roberts. Yet despite the large amount of research in the 1950s and 1960s that undermined their assumption that blood group alleles are adaptively neutral,[152] biology textbooks occasionally cited the Dunkers as an example of genetic drift (Elliott 1963, 1968).

152. Glass did acknowledge this later (1962). Another possible example of the founder/drift effect is in the Amish communities studied by Victor A. McKusick and his colleagues at Johns Hopkins; see McKusick (1978); Falconer (1981, 76).

17

....

Gould: Why Did the Synthesis Harden?
The Changing Views of
Dobzhansky and Wright

In 1983 Stephen Jay Gould extended his complaint about hardening beyond Simpson to Dobzhansky and Wright, the two founders of the synthesis who had tempered natural selection by genetic drift within small isolated populations in their explanations of speciation. He quoted from the first (1937) and third (1951) editions of *Genetics and the Origin of Species* to show "increasing emphasis on selection and adaptation." As for Wright, he now denies that he had ever "advocated a radically non-Darwinian approach to evolutionary change by demoting selection and adaptation in favor of accident." On the contrary, his shifting-balance theory, "which does specify an important role for genetic drift, is strongly adaptationist— but that adaptation arises at a level higher than the traditional Darwinian focus on individuals." Gould accepts this interpretation of Wright's present theory, although it assumes a type of "hierarchical thinking" unfamiliar and uncongenial to most evolutionists (Gould 1983, 85–86) But he notes that William Provine, in research for his forthcoming biography of Wright, found that Wright has significantly changed his own views since 1930 in a pro-adaptationist direction:

> The careful reader in 1932 would almost certainly conclude that Wright believed nonadaptive random drift was a primary mechanism in the origin of races, subspecies, and perhaps genera. Wright's more recent view that the shifting balance theory should lead to adaptive responses at least by the subspecies level is found nowhere in the 1931 and 1932 papers.[153]

153. Gould, quoting Provine (1983, 86).

....

Provine supported Gould's hardening thesis in his article about Wright's theory in the same 1983 book (Provine 1983). His biography of Wright, published in 1986, documented the change of views in more detail, noted Wright's refusal to acknowledge that change, and described the resulting confusion. The primary reason why both Wright and Dobzhansky abandoned drift in favor of selection was the result of their joint research on *Drosophila pseudoobscura*, in which they were able to duplicate the cyclic changes in inversion frequencies found in natural populations by using the population box technique pioneered by L'Héritier and Teissier. "Working with Dobzhansky had caused Wright to reject random drift as a major factor in *D. pseudoobscura* while at exactly the same time Dobzhansky's *Genetics and the Origin of Species* (in its first two editions) emphasized random drift" (Provine 1986, 403).

Leaving aside Huxley and Mayr for the moment, we must ask: why did Dobzhansky and Wright move from a selectionist theory incorporating random drift to a purely or primarily selectionist theory (what I call the natural selection hypothesis) in the 1940s? Their switch surely has a lot to do with explaining why other leaders and supporters of the synthesis began to favor a hard version.

Gould discussed Dobzhansky's shift toward strong selectionism in his introduction to the 1982 reprint of the first edition (1937) of *Genetics and the Origin of Species*. Comparison of this with the third edition (1951) showed that, like Simpson and Wright, Dobzhansky had hardened his theory by deleting

> the two chapters that contained most material on nonadaptive or non-selected change (polyploidy and chromosomal changes, though he includes their material, in reduced form, within other chapters). He adds a new chapter on "adaptive polymorphism" (1951, 108–34). He argues that anagenesis, or "progressive" evolution, works only through the optimizing, winnowing agency of selection based on competitive deaths ... But the most remarkable addition occurs right at the beginning. I label it remarkable because I doubt that Dobzhansky really believed what he literally said: I feel sure that he would have retracted or modified it had anyone pointed out that he had allowed a fascination for adaptationism to displace the oldest of evolutionary truths.
>
> In this addition ... Dobzhansky poses the key question of why morphological space is so "clumped":—why a cluster of so many cats, another of dogs, a third of bears, and so much unoccupied morphological space between? (Gould 1982)

Gould went on to accuse Dobzhansky (who died in 1975) of misusing Wright's "adaptive landscape" model to support adaptationism and rejects as "skewed" and irrelevant to evolution his explanation of "clumping." After two pages devoted to refuting Dobzhansky's new argument for selection, Gould offered only one paragraph to explain why Dobzhansky, Simpson, and others had become more favorable to the natural selection hypothesis since 1940:

> I do not fully understand why this hardening occurred, but I regard it as an important topic for historical research since its result so dominated the research program of evolutionary biology for many years. In part, the "ecological genetics" of E. B. Ford and his panselectionist school in England must have had a major effect. Their commitment to adaptationist explanations of everything and their discovery of strong selection coefficients in nature buoyed strict Darwinian faith. Dobzhansky must also have gained confidence from the claims of macroevolutionary colleagues like Simpson, Rensch, and Schmalhausen, who held that selection and adaptation might triumph in the very realm that had once denied it most strongly. Dobzhansky's 1951 argument for continuity between micro- and macroevolution (p. 17) is much stronger than the 1937 version. Finally, Dobzhansky's own empirical work increased his belief in the power of selection. In 1937, he tended to attribute inversion frequencies in natural populations of *Drosophila* to genetic drift, but he then discovered that these frequencies fluctuate in a regular and repeatable way from season to season, and decided (with evident justice) that they must be selective.

In the epilogue to his introduction, Gould stressed Dobzhansky's background as a "trained taxonomist and entomologist," which allowed him to perceive the evolutionary significance of certain peculiar mutations of *Drosophila*. I infer from Gould's remarks the suggestion that Dobzhansky gave greater weight to his own results in the area of his "first love"—entomology—than to evidence from other organisms (Gould 1982). Indeed, the index to the 1951 edition of *Genetics and the Origin of Species* provides prima facie evidence for this suggestion: there are more than one hundred references to *Drosophila*, compared to four to E. B. Ford, twelve to G. G. Simpson, thirteen to B. Rensch, and nine to I. I. Schmalhausen. But Dobzhansky also makes it clear that the theoretical interpretation of his *Drosophila* experiments is based on "the theory of balanced polymorphism ... developed by Haldane, Fisher, Ford, Wright, and others" (Dobzhansky 1951, 116).

By 1957, Dobzhansky, contrary to what one might expect from Gould's account, was launching his own campaign against the hardening of the synthesis. In a paper with Olga Pavlovsky, he attacked the view that natural selection is all-important, while drift can be ignored.

> As soon as a gene is shown to have any effect whatever on fitness, the conclusion is drawn that its distribution in populations must be determined solely by selection and cannot be influenced by random drift. But this is a logical non-sequitur. The important work of Aird et al. (1954) and of Clarke et al. (1956) disclosed that the incidence of certain types of gastro-intestinal ulceration is significantly different in persons with different blood groups. This is, however, far from a convincing demonstration that the observed diversity in the frequencies of the blood group genes in human populations is governed wholly, or even partially, by selection for resistance to ulcers. To make such a conclusion tenable it would have to be demonstrated that the environments in which human racial differences have evolved actually favored greater resistance in certain parts of the world and lesser resistance in certain other parts. Thus far no evidence has been adduced to substantiate any such claim. (Dobzhansky and Pavlovsky 1957, 311)

In a carefully designed experiment, Dobzhansky and Pavlovsky tested the prediction that because of genetic drift, a small population will evolve with greater genetic diversity than a large one starting with the same frequencies, even when both are subject to strong selective forces. It was also a test of Mayr's founder principle. Ten small populations (twenty founders each) and ten large populations (4000 founders each) of *D. pseudoobscura* were constructed to have a 50:50 ratio of *PP* (Pikes Peak) and *AR* (Arrowhead) gene arrangements in their third chromosome, and they were kept in population cages under uniform conditions for eighteen months (about nineteen generations). The results showed that the small populations had significantly greater variation in the *PP* and *AR* frequencies. "Although the trait studied . . . is subject to powerful selection pressure," as found in Dobzhansky's earlier experiments (chapter 12), "the outcome of the selection in the experimental populations is conditioned by random genetic drift. The either-selection-or-drift point of view is a fallacy" (Dobzhansky and Pavlovsky 1957, 316). Mayr's founder principle was also verified by the experiment.

The Dobzhansky-Pavlovsky experiment is regarded as a classic demonstration of the phenomenon of genetic drift,[154] and it certainly shows that

154. Brousseau (1967); Ruse (1981, 32–33); and Reisman and Forber (2005).

. . . .

predictions from the Fisher-Haldane-Wright theory can be tested under controlled conditions. But does it have any relevance to evolution in nature? Dobzhansky and Pavlovsky argued that it explains the fact that populations of the butterfly *Maniola jurtina*, "rather uniform throughout Southern England, despite some obvious environmental diversity in different parts of the territory[,] ... show quite appreciable divergence on the islands of the Scilly archipelago, although these islands are within only a few miles of each other and their environments appear rather uniform" (Dobzhansky and Pavlovsky 1957, 317). This divergence was attributed entirely to selection by Dowdeswell and Ford (1953) and Ford (1964, 54–58). Dobzhansky withdrew his earlier statements (1937, 1941) that the situation arose "through random drift in populations of continuously small size, or frequently passing through narrow 'bottlenecks.'" Instead, "in the island population we are observing the emergence of novel genetic systems moulded by interaction of random drift with natural selection" (317).

But Ford strongly rejected the view that his results from Scilly could be ascribed to the founder principle or any other kind of drift, insisting that they are completely due to natural selection (Ford 1975, 66–75).

Dobzhansky's change from pro-drift to anti-drift has already been discussed at length by Beatty, Smocovitis, Gould, and Provine (see chapter 12). It remains to note only that by 1970, he was beginning to suspect that it was time to reverse his position again, in the light of recent "theories of non-Darwinian evolution by random walk." He characterized these theories, which were to become more popular in the 1970s, as "a new swing of the pendulum in the opposite direction" (Dobzhansky 1970, 262). He told Jeffrey Powell in 1974, "I began as a drifter and then became a selectionist. Now, in my old age, I find myself becoming a drifter again" (Powell 1987, 34).

What is this pendulum? In the 1930s it pointed to "Wright's random drift idea [which] appeared to offer a ready explanation of adaptively neutral changes"; twenty-five years later, it swung back so that "hyperselectionism became fashionable. All differences between populations and species were assumed to be products of natural selection" (Dobzhansky 1970, 262), as illustrated by Mayr's (1963) position. Now it was suggesting randomness once more. I will reserve my own speculative answer for the concluding chapter.

During this period, Wright found himself in the awkward position of having to defend his views on the evolutionary role of isolation and genetic drift against the attacks of Fisher and Ford, while at the same time appearing to retract or drastically modify those views. In the famous Fisher versus Wright controversy, the Fisher side is often portrayed as selectionist and the Wright side as anti-selectionist. During the period up to 1970, Fisher seemed to be winning the battle, while Wright was either losing or claiming that he was *not* anti-selectionist. At the same time, the perception that there *was* a battle seemed to stimulate fruitful research and thus advanced the science while confusing the history (Provine 1985; Skipper 2000). In any case, by 1950 Wright was no longer opposing the hardening of the evolutionary synthesis.

Beginning with Maxime Lamotte's work on *C. nemoralis* in the 1950s (Lamotte 1950, 1959; Millstein 2007b), new evidence seemed to support genetic drift, yet Wright did not accept this evidence as support for his own theory; in fact, he "admitted that two of the prime examples of random drift he had frequently cited were actually cases of adaptations resulting primarily from selection."[155] But he continued to argue that genetic drift could occur in the laboratory as theoretically predicted and that it could act in combination with natural selection in evolution. While objecting to the premise that there is any "single cause for evolution," he stated that "in so far as my position could be classified as advocacy of any single principle, this principle was selection" (Wright 1955, 16). Later he asserted that he had never "attributed any evolutionary significance to random drift except as a trigger that may release selection" (Wright 1967, 255), although Provine (1986) has shown that Wright's recollection of his own early writings was not always accurate.

At the same time, the credibility of Mayr's founder principle did not seem to suffer from the refutations of genetic drift by the Ford group. While Wright and others considered the founder principle a special case of genetic drift, it came to be regarded as a distinct hypothesis, perhaps the only exception to the pro-selectionist consensus circa 1960.[156]

Gould did not explain why he himself, in a 1966 paper, expressed a strongly selectionist view: "Our ultimate goal in the study of a phyletic lineage is the explanation of each morphological change in terms of its selective advantage" (Gould 1966, 621). But we have ample evidence from the statements of other leading evolutionists to suggest why they dismissed the evolutionary significance of drift. As Timothy Shanahan suggests, "A number of celebrated cases of 'nonadaptive evolution by random drift' were later successfully reinterpreted as adaptations formed by natural selection." (Shanahan 2004, 134).

155. Provine (1986, 443); see Wright (1951, 455).
156. Provine (1986, 407–8, 484); Mayr (1963, 209–11, 529–34); and Mather (1973, 84).

18

....

Why Did the Synthesis Harden?
The Views of Other
Founders and Leaders

To begin with the founding fathers: Fisher, as far as I know, never deviated from his strong selectionist position.[157] Wright more or less abandoned his drift theory, while others adopted what they thought was his theory. Haldane, in the early 1950s, seemed to favor Wright's earlier view that drift assists selection.[158] Later he asserted that "chance effects . . . will rarely matter to a whole species," except in the case of very small populations where the founder principle may come into play or "the simultaneous establishment of several factors which are harmful singly but adaptive in combination" (Haldane 1958, 17). In a review of the then present status of natural selection for one of the many publications celebrating the centennial of Darwin's *Origin of Species*, Haldane briefly mentioned the idea that evolution may result from genetic drift in small tribes. He wrote, "The weakest point in Wright's argument is that he has not adequately considered what happens when this tribe starts hybridizing with others" (Haldane 1959, 141).

Ford, like his colleague Fisher, always insisted that genetic drift has no important role in evolution but went further in his public criticism of Wright: the theory of adaptive peaks

> is wholly unrealistic and has done much harm in biology; assuming, as it apparently does, that ecological conditions are so stable that a given type of genetic adaptation . . . is persistent. (Ford 1957a, 86)

157. Bennett (1983, 46); see also Fisher (1950, 19) and Fisher and Ford (1947, 167–71; 1950).
158. Haldane (1953, foreword; 1954, 117). The first comment simply endorses Waddington's statement (1953, 186) that genetic drift is one of the "very few qualitatively new ideas" that have emerged from the mathematical theory of evolution.

He admitted that when the theory of genetic drift was proposed, it seemed reasonable because selective advantages were thought to be small; but we now know, Ford insisted, that "large selective advantages are in fact common in natural populations," so selection overwhelms drift.[159] The verdict: experiments by Ford and his colleagues proved that natural selection explains everything, while drift explains nothing.[160]

Julian Huxley, an influential advocate of the synthesis, did not show much sympathy for genetic drift after 1942. Returning from a two-week visit to the Union of Soviet Social Republics in 1945, Huxley reported with approval that despite Lysenko's influence, Russian evolutionists were even more selectionist than those in Britain and the United States and had kept their pure research going more effectively, despite wartime pressure to do applied work. He mentioned Nicolai Dubinin's work on seasonal selective changes and control of mutations and Gause's "first experimental proof of the efficacy of organic selection" using ciliates.[161] Reviewing Simpson's *Tempo and Mode in Evolution*, Huxley noted that gaps in the fossil record might be explained away by Wright's hypothesis that evolution is more rapid in small isolated populations, but geneticists may doubt if such populations could exist for the millions of years needed for nonadaptive evolution (Huxley 1945a). Later he wrote, "It is now clear that selective advantages so small as to be undetectable in any one generation, are capable, when operating on the scale of geological time, of producing all the observed phenomena of biological evolution. . . . Evolutionary change is almost always gradual and is almost wholly effected through selection" (Huxley 1954, 3). He asserted that the discovery that human blood groups are maintained in a morphism by a balance of selective advantages and disadvantages (chapter 13) confirms Fisher and refutes Wright's suggestion that the ratios originated by drift (Huxley 1955). In later writings he simply omitted genetic drift as a significant factor in evolution, until 1963, when he again changed his mind and stated that drift "can have definite evolutionary consequences."[162]

In an abridged edition of his 1963 book published seven years later, Mayr took notice of the proposed revival of genetic drift under the name "non-Darwinian evolution" (inappropriate and misleading, he asserted, because Lamarckism, orthogenesis, and other heresies could also be described

159. Ford (1957a); see also Ford (1971, 36–38).
160. Dowdeswell and Ford (1953) and Ford (1955, 1960).
161. Huxley (1945b, 255); cf. Kolchinsky (2000a).
162. Huxley (1963, introduction) and Olby (1992, 70).

by that term). He attempted to refute the new ideas in a couple of pages, concluding that they should not replace natural selection as the primary cause of evolution (Mayr 1970, 126–8).

Stebbins asserted that the burden of proof lies on the shoulders of anyone who wants to invoke genetic drift to explain nonadaptive differences: "If we cannot see why a difference should be connected with differential adaptation, we are not justified in concluding from this fact that the difference is non-adaptive" (Stebbins 1966, 74). Contrary to early assumptions that genetic drift should be relatively more important in small populations,

> no example of differentiation between populations is known which can be ascribed solely or principally to [chance]. One reason is that the effects of selection, both direct and indirect, are very strong, particularly in populations which are becoming smaller because of a worsening environment. Consequently, no gene remains unaffected by selection for a sufficient number of generations so that its frequency can be permanently altered by chance alone. (Stebbins 1966, 76)

Many phenomena *can* be explained by a *combination* of chance and natural selection. However, "Chance factors by themselves probably have little effect on evolutionary processes, but in combination with reduction in population size and the accompanying natural selection, they may play an important role in guiding the early stages of adaptive differentiation" (Stebbins 1966, 77). This is essentially Wright's later view. But three years later, Stebbins restated Wright's concept of adaptive peaks and valleys by proposing that mutations, rather than drift, trigger the jump from one peak to another (Stebbins 1969, 124–5). Presumably he meant small mutations. This was a nice balancing act on the part of Stebbins, since he insisted at the same time that the Goldschmidt macromutation theory had been definitely excluded by modern molecular biology and developmental genetics (Stebbins 1969, 104).

Stebbins also had a notion of fashions in evolutionary theory, which he did not (as far as I know) mention in any of his scientific publications but described in a letter to Edward O. Dodson, who summarized it in his 1952 textbook. According to Stebbins, there were three periods of evolutionary thought after the *Origin of Species*: (1) a romantic period (1860 to 1903), characterized by "extreme enthusiasm for Darwinism and adaptationism"; (2) an agnostic reaction (1903 to 1935), in which the rediscovery of Mendel's laws and the invention of the gene concept seemed to lead to a dead end because of Johannsen's conclusion that natural selection has a limited effect in improving a pure line and because of de Vries's unwieldy large mutations; and (3) the modern synthesis, beginning around 1935, in which Darwinism was revived.[163]

Why did the synthesis harden? David J. Depew and Bruce H. Weber write:

> Gould professes not to know the causes of this "hardening." He mentions the prestige of solid adaptationist explanations in the field (Gould 1983, 88). These seemed to confirm selectionist experiments conducted in the laboratory, where gene frequencies of different phenotypes could slowly be shifted by controlled environmental variation. We would like to suggest, however, that standing behind this prestige is the deepening influence of empiricist models of science on the synthesists, their successors, and their philosophical defenders. A clean series of inferences linking a mathematically expressible set of laws to laboratory work, and extending laboratory results to the field, was precisely what would be needed to argue that evolutionary theory now stood on a solid, quantifiable, and potentially axiomatic-deductive basis. (Depew and Weber 1985b, 230)

163. Dodson (1952, 91–94); see also Cockrum and McCauley (1965, 631).

19

....

The Peppered Moth

In addition to the evidence presented by Dobzhansky (1941) and summarized in chapter 7, and the more recent research on chromosome inversions (chapter 12), blood groups (chapter 13), resistance to antibiotics (chapter 14), and snails and tiger moths (chapter 15), many other observations and experiments were cited in books published during the four decades covered by this survey. I will mention here only one, which became quite well known in the 1960s.

Industrial melanism was frequently mentioned as a probable example of natural selection, even before the research by H. B. D. Kettlewell in England on the peppered moth (*Biston betularia*).[164] It was widely noted in the late nineteenth century that in industrial areas of England and other countries, moths that had previously had light-colored wings were being replaced by moths (of the same species) with dark-colored wings. It seemed plausible that this phenomenon was somehow connected with the darkening of tree bark caused by smoke from factories, but there was considerable disagreement about the exact nature of the selective factor. E. B. Ford (1937, 1945) argued that the dark (melanic) form was more viable, but its numbers were ordinarily limited because dark moths were more likely to be seen and eaten by birds when the trees were a lighter color. Others suggested a more direct connection between the chemical compounds in polluted air and the physiological mechanism that determined wing color.

Starting in 1953, Kettlewell, a member of Ford's group at Oxford, tested the hypothesis that the moths were being eaten by birds and that they were more likely to be eaten when the color of their wings contrasted with the color of the trees on which they rested. He marked and released more than 200 moths, mostly the dark form (*carbonaria*), in a heavily polluted area near Birmingham and then tried to recapture them. A significantly higher proportion of the light moths disappeared, and several were observed being taken by robins and sparrows. Niko Tinbergen, a Dutch-British biologist

---

164. On Kettlewell, see Turner (1990) and Rudge (2003, 2005, 2006).

....

who later received the 1973 Nobel Prize in Medicine or Physiology for his research on animal behavior, took motion pictures of redstart birds taking and eating the moths. As a control, the experiment was repeated in the unpolluted countryside of Dorset. There, as expected, the dark form was much more likely to escape the birds. Kettlewell claimed to have established not only that predation by birds was the selective agent, but also that human judgments of which moths were more conspicuous against a light or dark background correlated well with the effectiveness of the birds in catching the moths.

Kettlewell's first results were published in 1955, with a full report in 1956 and follow-up papers in later years. A 1959 article in *Scientific American* probably made his work better known to biologists and teachers in the United States. As one might expect, the 1955 report was immediately cited by Ford (1955) at a Cold Spring Harbor colloquium, where he challenged the Americans to study industrial melanism in their own country. During the next few years, it was mentioned in books by Ford's colleagues (Dowdeswell and Sheppard) and by other British biologists.[165] Kettlewell's experiment did not appear in any American book on evolution, as far as I know, until 1961. In that year, Bruce Wallace and Adrian M. Srb (both at Cornell University) mentioned it briefly in their book *Adaptation* (Wallace and Srb 1961, 40). Since this book appeared in a second edition in 1964, we may assume it was widely read and was thus partly responsible for publicizing Kettlewell's work. The next year it was noted in four books on evolution by American authors: Eldon John Gardner at Utah State University, David J. Merrell at the University of Minnesota, Paul Amos Moody at the University of Vermont, and Herbert H. Ross at the University of Illinois.[166]

The earliest reference I have found in a zoology text is by Alfred M. Elliott at the University of Michigan. He and Charles Ray Jr. at Emory University seem to have been the first to mention it in a general biology text (Elliott and Ray 1965); there is also a vague reference to the changes in populations of light and dark moths by G. G. Simpson and William S. Beck, both at Harvard.[167] After that, it became a familiar piece of evidence for natural selection in many textbooks, most of which reproduced Kettlewell's pair of photographs showing moths against similar and contrasting backgrounds.

165. Lack (1957); Dowdeswell (1958, 1959, 1963); Sheppard (1959b); Barnett (1958); Waddington (1961); de Beer (1962); and Huxley (1963).

166. Gardner (1962); Merrell (1962); Moody (1962); and Ross (1962).

167. Elliott (1963); Elliott and Ray (1965); and Simpson and Beck (1965).

Did Kettlewell's research on the peppered moth help to persuade evolutionists to accept natural selection as the primary cause of evolution? No. Most had already adopted that position, or like Moody (1953, 1962), continued to insist on a significant role for genetic drift in evolution. But through its appearance in many textbooks, popular articles, and even a novel by Margaret Drabble, it may have reinforced the evolutionary beliefs of the next generation of biology students[168]—although by that time, the evidence for evolution and for the natural selection hypothesis was already overwhelming.

168. Drabble (2001, title page, 252); Hagen (1999); and Rudge (2000b).

20

············

The Triumph of
Natural Selection?

According to Ernst Mayr, at a meeting in Princeton in January 1947,

> there was universal and unanimous agreement with the conclusions of the
> synthesis. All participants endorsed the gradualness of evolution, the pre-
> eminent importance of natural selection, and the populational aspect of
> the origin of diversity.... Not all other biologists were completely con-
> verted. This is evident from the great efforts made by Fisher, Haldane,
> and Muller as late as the late 1940s and 50s to present again and again
> evidence in favor of the universality of natural selection, and from some
> reasonably agnostic statements on evolution made by a few leading biolo-
> gists such as Max Hartman. (Mayr 1982, 568–9)

By the 1960s, the founders and leaders of the evolutionary synthesis in the
United States and England (table 1, p. 131) had decided that random genetic
drift by itself, while theoretically expected to occur under certain conditions,
has essentially no significant role in evolution, except perhaps to facilitate
natural selection, so that natural selection acting on small variations is all-
important. The views of founders and leaders have been discussed in earlier
chapters. I now turn to four other biologists whose work was often cited as
evidence for natural selection: N. P. Dubinin at the Institute of Cytology
in Moscow, Francis B. Sumner at the Scripps Institute of Oceanography
in La Jolla, W. C. Allee at the University of Chicago, and David L. Lack
at Oxford.

Dubinin, who had earlier proposed an idea similar to Wright's genetic
drift, started to publish his research in western journals in 1945. In a series
of papers with G. G. Tiniakov, he reported that natural selection rapidly
produces measurable changes in the genetic structure of populations of

············

Drosophila funebris near Moscow. No mention of genetic drift (or Dubinin's own "genetic automatic processes") appears in these papers.[169] The selective factors in them are seasonal changes in climate and industrialization.

Sumner, whose work on protective coloration of the deer mouse (*Peromyscus*) and the mosquito fish (*Gambusia patruelis*) was often cited as evidence for natural selection, was originally a Lamarckian and also contemplated the possibility of nonadaptive evolution by drift. He was persuaded by his own results to accept an adaptationist interpretation. Since Lamarckism had been refuted and "inner perfecting tendencies [orthogenesis] do not belong in the realm of science," natural selection was the only hypothesis worth considering. Goldschmidt's (1940) revival of macromutation was unacceptable—as Darwin wrote, it amounted to believing in miracles (Sumner 1942, 437). In his autobiography, Sumner wrote that he decided the color of the mice was a result of selection by predators, in accordance with Mendelian laws, despite his dislike for particulate theories in science and for the faddish character of the Mendelian movement (Sumner 1945, 239–43). "Because he enjoyed a well-deserved reputation as an exceedingly careful experimentalist, his reinterpretation of the adaptive value of the differences between geographical races carried much . . . weight with both systematists and geneticists."[170]

Allee, who followed Wright's theory of evolution in his 1938 book on the social life of animals, published (with four other authors) a massive treatise on animal ecology in 1949. Although he still seemed to favor Wright's theory, it was now overshadowed by a huge amount of evidence for natural selection (Allee et al. 1949).

Lack, an ornithologist known for his research on Darwin's finches, is an example of a scientist who, like Dobzhansky, changed from drift to natural selection as a result of his own research.[171] In his monograph on the Galapagos finches, Lack rejected Darwin's view that natural selection would favor different varieties on different islands, because he could find no evidence that the differences have adaptive significance. Instead, like "most geographical forms which have been described in birds," their nonadaptive evolution as "small and isolated populations" was best described by the Sewall Wright effect (Lack 1945, 116–7). In an autobiographical memoir, Lack wrote:

169. Dubinin and Tiniakov (1945, 1946a, 1946b, 1946c, 1947a, 1947b).
170. Provine (1986, 230; see also 1979) and Gerson (1998, 363).
171. Provine (1986, 406–8) and Gould (1983, 88).

Since . . . Arthur Cain asked me, in the mid-1950s, why I postulated that various subspecific differences in the finches are nonadaptive, it is worth stressing that, before my work, almost all subspecific differences in animals were regarded as non-adaptive (hence the importance of Sewall Wright's theory of genetic drift). . . . "The variation of animals in nature" by G. C. Robson and O. W. Richards in 1936 fairly reflected current opinion. . . . I reached my conclusion, that subspecific and specific differences in Darwin's finches are adaptive, and that ecological isolation is essential for the persistence [*sic*] of new species, only when reconsidering my observations five years after I was in the Galapagos. (Lack 1973, 429)

In the first edition of his book *Darwin's Finches,* Lack stated that, contrary to Robson and Richards, he believed "the absence of adaptive differences [between related species] is only apparent"; adaptive differences must exist. "There is no other way of accounting for the ecological isolation of each species" (Lack 1947, 134, 142). He was already more selectionist than Huxley, who in his 1942 book had stated that "ecological isolation is a cause of species-formation," whereas Lack asserted that speciation is "equally explicable through competition between species" (Lack 1947, 142).

When *Darwin's Finches* was reprinted in 1961, Lack had become director of the Edward Grey Institute of Field Ornithology at Oxford, but the publisher would not let him change the text, so he could only add a preface, in which he explained that the burden of proof was now on anyone who claimed that species differences are nonadaptive:

This text was completed in 1944. . . . In the interval, views on species-formation have advanced. In particular, it was generally believed when I wrote the book that, in animals, nearly all of the differences between subspecies of the same species, and between closely related species in the same genus, were without adaptive significance. . . . Sixteen years later, it is generally believed that all, or almost all, subspecific and specific differences are adaptive, a change of view which the present book may have helped to bring about. Hence it now seems probable that at least most of the seemingly non-adaptive differences in Darwin's finches. . . would, if more were known, prove to be adaptive. (Lack, 1961)

In his book *Ecological Adaptations*, Lack did not mention drift, but wrote:

This book, like every other work of natural history, shows, at least for those who have eyes to see, the immense power of natural selection and the complex and subtle adaptations to which it can give rise. (Lack 1968, 310)

Lack's 1947 book had an immediate influence on one biologist. Alistair Cameron Crombie (who later became a historian of science) wrote in the *Journal of Animal Ecology* that the doctrine of Robson and Richards (1936)—that species differences are nonadaptive—could now be rejected, since Lack had shown the differences between bird species to be adaptive (Crombie 1947).

Although the research by Lack, and later by Peter and Rosemary Grant, contributed to the general acceptance of Darwinian evolution, their influence came primarily after 1965, when the NSH was already established. Edward J. Larson's comprehensive book on this subject states that "Lack's view of the process [of speciation] was textbook dogma by the 1960s" but cites only one book published before 1965 to support that statement.[172]

Results of a Survey of Biological Publications

Given this consensus of the founders and leaders of the evolutionary synthesis in favor of natural selection—denying any significant role for genetic drift or macromutations—what about the "followers"? —That is, what about other biologists who published books and technical review articles on evolution and related topics during the 1940s, 1950s, and 1960s? Here, the picture is not as clear-cut.

In the decade 1941–1950, less than 40 percent of those publications accepted the NSH (natural selection acting on small mutations) as the primary factor in evolution and indicated that drift is insignificant or did not mention it at all (see table 3 [p. 133] for details). Almost 30 percent, while accepting natural selection as an important factor, stipulated that drift may also have a significant influence on evolution. Some would also allow a role for other factors: large mutations (13 percent) and/or orthogenesis (13 percent). These numbers add up to more than one hundred percent because the categories are not exclusive; for example, Breder (1944) was willing to accept both large mutations and orthogenesis. About 13 percent rejected natural selection entirely.[173]

172. This was a high school textbook (American Institute of Biological Sciences, Biological Sciences Curriculum Study [1963]). The quote is from Larson (2001, 198); references are in his note 6 (295–6). Most of the high school texts published before 1961 did not mention evolution at all (Grabiner and Miller 1974).

173. Natural selection without drift (publications): 1941: Hubbs; 1942: none; 1943: Ashby, Emerson, Hubbs (2), Mather, Willier; 1944: Conklin, Wigan; 1945: Burnet, Clausen et al., White; 1946: none; 1947: Lull, Luria; 1948: none; 1949: Colbert, Jepsen; 1950: Summerhoff.

Natural selection with drift: 1941: Dice; 1942: none; 1943: Amadon, Hazel; 1944–1945: none; 1946: Lush; 1947: Crombie; 1948: Li, Muller, Schaeffer; 1949: Allee et al., Boyd, Muller; 1950: Birdsell, Boyd, Montagu.

In the next decade (1951–1960), about 60 percent accepted natural selection acting on small mutations, without drift, as the primary or only factor in evolution. Support for drift (along with natural selection) remained just below 30 percent.[174]

In the last decade surveyed, 1961–1970, support for natural selection without drift dropped slightly but remained about 50 percent, while support for drift increased to about 35 percent. By the second half of this decade (1966–1970), large mutations, Lamarckism, and orthogenesis had completely disappeared.[175] So, the views of these other evolutionary biologists in the late 1960s reflect the views held by the founders and leaders

---

Large mutations: 1941: none; 1942: Breder; 1943: Westall; 1944: none; 1945–1946: Goldschmidt; 1947–1948: none; 1949: Davis, Lamy; 1950: none.

Lamarckian: 1941–1950: none.

Orthogenesis: 1941: none; 1942: Breder; 1943: none; 1944: Conklin; 1945: Lillie; 1946: none; 1947: Lull; 1948: Holmes; 1949: none; 1950: Summerhoff.

Rejects natural selection: 1941: none; 1942: Breder; 1943: McAtee, Willis; 1944: Conklin; 1945: Lillie; 1946: Errington; 1947–1950: none.

174. Natural selection without drift (publications): 1951: Blum, Clausen, de Beer, Gregory, Rice and Andrews, Watson; 1952: Mathews; 1953: none; 1954: Cain, Lerner, Sheppard, White; 1955: none; 1956: Harvey; 1957: Carter; 1958: Clausen and Hiesey, Darlington, Dowdeswell, Kettlewell, Sheppard; 1959: Anfinsen, Clark, Dowdeswell, Lerner, Morley, Sheppard (2); 1960: Birch, Sheppard, Waddington.

Natural selection with drift: 1951: Carter, Shull; 1952: Dodson, Lindsey, Patterson and Stone; 1953: Moody; 1954: Carter; 1955: Li; 1956: none; 1957: Waddington; 1958: J. M. Smith; 1959: Dunn, Lamotte; 1960: Van Valen.

Large mutations: 1951: none; 1952: Dodson; 1953–1954: none; 1955: Goldschmidt; 1956: none; 1957: Waddington; 1958–1959: none; 1960: Dodson.

Lamarckian: 1951: none; 1952: Lindsey; 1953: none; 1954: Carter, Russell (published 1962); 1955–1956: none; 1957: Carter, Waddington; 1958: Cannon; 1959–1960: none.

Orthogenesis: 1951: Carter, de Beer; 1952–1953: none; 1954: Carter; 1955–1957: none; 1958: Brough, de Beer; 1959–1960: none.

Rejects natural selection: 1951–1952: none; 1953: none; 1954: none; Heuts; 1954–1957: none; 1958: Cannon; 1959–1960: none.

175. Natural selection without drift: 1961: Lack, Wallace and Srb; 1962: de Beer, Gardner; 1963: Cain, Darlington, Dowdeswell; 1964: none; 1965: Blacklith, Hardy, Levins, MacArthur; 1966: MacArthur and Connell, Ross, Solbrig, Williams; 1967: Hamilton, Sheppard; 1968: Allison, D. L. Clark, Gottlieb, Workman; 1969: Ewens; 1970: none.

Natural selection with drift: 1962: Merrell, Moody; 1963: Ehrlich and Holm; 1965: Carlquist; 1966: J. M. Smith; 1967: Dunn; 1968: none; 1969: Mettler and Gregg, Weller; 1970: Crow and Kimura, Moody, Volpe.

Large mutations: 1951–1960: none; 1962: Ross; 1963–1970: none.

Lamarckian: 1951–1960: none; 1961: Waddington; 1962–1964: none; 1965: Hardy; 1965–1970: none.

Orthogenesis: 1951–1962: none; 1963: Sinnott; 1964–1970: none.

Reject natural selection: 1951–1960: none; 1961: Waddington; 1962: Blackwelder; 1963: Sinnott; 1964–1966: none; 1967: Carter; 1968: Cronquist; 1969–1970: none.

. . . .

in the 1940s after rejecting "other factors" but before the hardening of the synthesis (rejection of drift).

Another group of publications to consider here includes textbooks, popular books, and articles (again excluding those written by founders and leaders). Although textbooks are often dismissed by scientists because they are not written by active researchers and do not cover the latest results, historians have recently begun to recognize their value in reflecting the accepted knowledge of a scientific community (Brush 2005, Bertomeu-Sánchez et al. 2006). It is important to learn what the leading scientists think about the validity of a theory, and most of the preceding chapters of this book have been devoted to that question. But ideas do not really become established knowledge in a scientific discipline until they are taught to the next generation. Textbooks may not report what the leaders think today, but they do tell us what students are learning, and since many college textbooks are written by college professors, they give an indication of what is accepted by academic scientists. Furthermore, the *changes* in the content of textbooks over time reflect, with some time lag, changes in the views of the leaders. Conversely, if an idea *never* appears in a textbook, it can hardly be said to be established knowledge.

The value of textbooks is succinctly expressed in a statement by professional scientist and amateur historian Stephen Jay Gould:

> To learn the unvarnished commitments of an age, one must turn to the textbooks that provide "straight stuff" for introductory students.... Surveys of textbooks provide our best guide to the central convictions of any era.... This field of vernacular expression has been neglected by scholars, though the subject would yield great insight. (Gould 2002, 576–7)

In my survey, I have included popular books and articles in the category of textbooks in order to get a reasonably large sample; a more comprehensive study would probably reveal significant differences.

In the decade 1941–1950, if we first disregard the question of drift, we find that whereas about two-thirds of the evolutionary biologists accepted natural selection acting on small mutations in their books and technical reviews, about half of the textbooks did so (see table 2, p. 132). Fewer than 15 percent of evolutionary biologists supported large mutations, compared to nearly 45 percent of authors of texts (see table 2). I would identify this decade as the time when evolutionary biologists as a community rejected de Vries (and Goldschmidt) and accepted the hypothesis that evolution proceeds by natural selection acting on *small* mutations, while many other biologists continued to support evolution by large mutations.

In the same decade, while drift was gaining support among evolutionary biologists (see table 3), it was gaining a foothold in textbooks but was not quite as popular as orthogenesis (see table 2).[176]

In the next decade, 1951–1960, the proportion of textbooks that endorsed natural selection (with or without drift) jumped from half to about three-quarters,[177] while the support for drift rose from 13 percent to about 20 percent.[178] Large mutations received the support of about 20 percent, considerably less than in the previous decade.[179]

Finally, in the decade 1961–1970, the textbooks were similar to the books and technical reviews in that they had almost 60 percent support for

176. Natural selection without drift: 1941: Colin, Fasten, Gerould and Poole; 1942: Rogers et al.; 1943: none; 1944: S. F. Cain, Howells; 1945: Borradaile, Kenoyer and Goddard, Marsland; 1946: Shull et al., J. A. Thomson; 1947: Emerson; 1948: Alexander, Holmes; 1949: Darlington and Mather, James, Nelson, Pauli; 1950: Baitsell, Bates, Etkin, Fuller, Gibbs, Grove and Newell, Moment, Robbins and Weir.

Natural selection with drift: 1941: Shull et al.; 1942–1944: none; 1945: Altenberg; 1946: Hogben, Sinnott; 1947: Curtis and Guthrie, Daubenmire; 1948: none; 1949: Hardin; 1950: Shull.

Large mutations: 1941: Fasten, Fuller, Guyer, Moon and Mann, O'Hanlon, Weymouth; 1942: Hunter and Hunter, Weatherwax; 1943: Baker and Mills, MacDougall and Hegner, Storer; 1944: Grove and Newell, Strausbaugh and Weimer; 1945: Lowson; 1946: Haupt, E. G. White; 1947: Emerson, Mavor, Weatherwax; 1948: Guyer, Potter; 1949: de Laubenfels, Fuller, Hunter and Hunter (2), Stauffer, Wheat and Fitzpatrick, Winchester; 1950: Villee.

Lamarckism: 1941: Fasten; 1942–1943: none; 1944: Howells, Strausbaugh and Weimer; 1945: Lowson; 1946–1947: none; 1948: Guyer; 1949–1950: none.

Orthogenesis: 1941: Guyer; 1942: Swingle; 1943: MacDougall and Hegner, Storer; 1944: Grove and Newell, Strausbaugh and Weimer; 1945–1947: none; 1948: Guyer; 1949: Hunter and Hunter, Mavor; 1950: none.

Rejects natural selection: 1941–1942: none; 1943: de Laubenfels; 1944–1947: none; 1948: Carter; 1949–1950: none.

177. Natural selection without drift: 1951: Hegner and Stiles; 1952: Elliott, Gardiner, Hardin, Milne and Milne, Wilson; 1953: Darlington, Grove and Newell, Kenoyer et al., Transeau et al.; 1954: Breneman, Whaley; 1955: Mainz, A. Morgan; 1956: Allison, Brown; 1957: Grove and Newell, Guthrie and Anderson, J. A. Moore, Winchester; 1958: Milne and Milne, Whaley et al.; 1959: Breneman, Brouwer, Hegner and Stiles, Kettlewell (article), Mavor, Milne and Milne.

178. Natural selection with drift: 1951: none; 1952: Rogers et al.; 1953: Moore; 1954: none; 1955: Hall and Moog; 1956: Alexander, Johnson et al., Wells and Wells; 1957: Snyder and David, Villee; 1958: Yapp/Borradaile; 1959: Hardin, Weisz, Zirkle; 1960: none.

Natural selection rejected: 1955: Sinnott; 1958: Beaver.

179. Large mutations: 1951: Fuller, Stanford, Weimer; 1952: Milne and Milne; 1953: Kenoyer et al., Transeau et al.; 1954: Fuller and Tippo, Villee; 1956: Haupt, Miller and Haub; 1957–1960: none.

Lamarckian: 1951–1953: none; 1954: Fuller and Tippo; 1958: Yapp/Borradaile.

Orthogenesis: 1953: Grove and Newell; 1957: Grove and Newell; 1960: Elliott and Ray.

Natural selection rejected: 1951–1954: none; 1955: Sinnott; 1956–1957: none; 1958: Beaver; 1959–1960: none.

natural selection without drift[180] and about 35 percent for natural selection with drift.[181] The major difference is that about 10 percent of the textbooks still endorsed large mutations,[182] which had almost disappeared from the monographs and technical reviews.

So, our conclusion—that the natural selection hypothesis became established knowledge in the 1960s—is just barely (and probably only temporarily) valid. The strong pro-selection consensus of the founders and leaders did not translate into equally strong support for natural selection among other biologists during the period of this study, a consequence of the familiar fact that knowledge presented in textbooks generally lags behind the research frontier by a decade or so.

One advantage of looking at textbooks popular enough to be frequently revised is that one can sometimes see an author changing his or her mind in successive editions and can thereby get an idea of what reasons were persuasive.[183] Here, we have four cases:

1. James Watt Mavor (1883–1963), professor of biology at Union College, changed from the de Vries large-mutation theory to the NSH (small mutations) between the second (1941) and third (1947) editions of his

180. Natural selection without drift: 1961: Bower, Brown, Core et al., Grove and Newell, Heiss and Lape, Winchester and Lowell; 1962: Beaver, Bonner, Buffaloe, Goodnight et al.; 1963: American Institute of Biological Sciences/BSCS, Braungart and Buddeke, Grogan, Hausman, Kenoyer et al., Stauffer et al.; 1964: W. B. Crow, Dillon and Dillon, Eisman and Tanzer, Goodnight and Gray, Guyer and Lane, Marsland; 1965: Miller and Vance, Nason, Penny and Waern, Storer and Usinger, Winchester; 1966: Avis, Berrill, Cockrum et al., Grove and Newell, Hardin, Romer, Stauffer et al., B. Wallace; 1967: Ashton, Baker and Allen, Buffaloe and Throneberry, Gardiner and Flemister, Keeton, Nelson et al., Platt and Reid, Wilson and Loomis.

181. Natural selection with drift: 1961: Hickman, Johnson et al.; 1962: Alexander, Villee; 1963: Elliott, Villee et al., Weisz, Yapp/Borradaile; 1964: Dillon and Dillon, Whaley et al.; 1965: Cockrum and MacCauley, Dillon, Weisz; 1966: Beaver and Noland, Berrill, Elliott and Ray, Johnson et al., Mavor and Manner, Weisz, Winchester; 1967: Moment, Villee, Weisz; 1968: Elliott, Lerner, Nason and Palmer, Storer et al., Strickberger, Villee et al., Weisz; 1969: Johnson et al. (general), Johnson et al. (zoology), Keeton, Orians, Weisz; 1970: Jessop, Nelson et al., Parsegian et al., Winchester.

182. Large mutations: 1961–1962: none; 1963: Villee et al.; 1964: Winchester; 1966: Avis, Mavor and Manner, Stauffer et al.; 1967: Villee; 1968: Taylor and Weber, Villee et al; 1969: Kroeber et al., Nason and Goldstein; 1970: none.

Lamarckian: 1961–1968: none; 1969: Orians; 1970: none.

Orthogenesis: 1961–1965: none; 1966: Mavor and Manner; 1967–1970: none.

183. According to "Planck's Principle," scientists (especially older, well-established ones) do not change their beliefs to accept a new idea; they just retire and die, while the younger generation, which has learned the new idea, takes over. This principle was first tested experimentally by David Hull, Peter D. Tessner, and Arthur M. Diamond, using Darwin's theory of evolution as the new idea (Hull et al. 1978). We have already seen that in the mid-twentieth century, several leaders did change their views about the importance of genetic drift in evolution.

General Biology text but forgot to remove his endorsement of de Vries in making this revision (Mavor 1947, 864, 880).

2. Alfred M. Elliott (1905–?), professor of zoology at the University of Michigan, was completely selectionist in the first and second editions of his zoology text (1952, 1957), but in later editions, he discussed the Dunkers (studied by Glass and his colleagues) as an example of genetic drift (Elliott 1963, 1968). At the same time, he gradually abandoned his support for orthogenesis, exemplified by the horse, which "seems to be a case of straight-line evolution, almost as if the horse 'wanted' to change its way of life from a shy, seclusive forest dweller to a swift running prairie animal" (1952, 658). In later editions, influenced by Simpson, he stated that the evolution was not as straight-line as it might seem (Elliott 1968). The successive editions show signs of inconsistent piecemeal revision, with contradictory statements on different pages.

3. William Beaver (1896–?), professor of biology at Wittenberg University, presented a negative view of natural selection in the fifth edition of his general biology textbook (1958), which he dropped in the sixth edition (1962), where he also mentions the "modern synthesis" (628). The seventh edition, coauthored by George B. Noland of the University of Dayton, gives a fuller treatment of evolution, stating that the major factors are (presumably in order of importance) mutation, genetic drift, natural selection, and migration (Beaver and Noland 1966, 506). No examples of natural selection are given.

4. One author presented one view in a popular book but then presented a different view in a textbook published a few years later. Garrett Hardin (1915–2003), professor at the University of California, Santa Barbara, discussed several cases in which genetic drift is important in *Nature and Man's Fate* (Hardin 1959). But in the second edition of his textbook on biology (Hardin 1966), he presented the theory of natural selection with no mention of drift.[184]

184. This is in effect the same as his position in the 1952 edition of the textbook (Hardin 1952). Hardin is best known for his influential paper, "The Tragedy of the Commons" (1968).

21

....

Is Evolutionary Theory Scientific?

Those biologists who attempted to revive the theory of natural selection in the 1930s were acutely aware that the theory had acquired a bad reputation because its earlier supporters were too eager to explain everything by a "just so story" like the fairy tales of Rudyard Kipling (1902). Raymond Pearl complained that "extremely few attempts have ever been made ... to test the principle of natural selection experimentally, or even observationally, by precise and critical *specific* observations," even though it is "perhaps the most important abstract biological principle ever enunciated. Tons of books and papers have been written about it" (Pearl 1930, 176). A. F. Shull, in agreement with Pearl, recommended that evolutionists, in their renewed efforts to revive the theory of natural selection, should forget about "its sexual selection, its warning colors, its mimicry and its signal colors" in favor of a more experimental approach (Shull 1936, 212). D. M. S. Watson, in the passage quoted in chapter 5, complained that the NSH is so flexible, it can explain anything. But two decades after Pearl's critique, substantial progress had been made in satisfying his requirements for "a proof that natural selection has altered a race" (Pearl 1930, 175), as shown by the detailed survey of Allee et al. (1949, 641–55).

Starting around the time of the 1959 centennial of Darwin's *Origin of Species*, philosophers and biologists began to debate the question, is evolutionary theory scientific? By this the philosophers meant, does it make testable predictions? The biologists had to decide whether to answer that question (as Haldane had already done in the quotation at the beginning of this monograph) or to argue that the philosophers' criterion should not be applied to evolutionary theory (or to biology in general). The question is especially relevant at the beginning of the twenty-first century: creationists attack Darwinism because it is supposedly not testable, and their attacks may seem plausible to nonscientists who have learned a naive version of the scientific method in high school science courses. Biology teachers (e.g., Cooper 2001) may try to defend evolution by arguing that reconstructing

....

past events, when no human was present to observe them, is a legitimate method in science (geologists and astronomers do it all the time), without realizing that evolutionists have in fact made testable (and confirmed) predictions. In fact, one of the most prominent creationists mentions Darwin's hypothesis about wingless insects as an example of an untestable tautological proposition (Johnson 1993, 21); he is apparently unaware of the experiment described in chapter 7.

Even before 1959, there were two important experiments that confirmed predictions of the Fisher-Haldane-Wright theory under controlled conditions, where both natural selection and genetic drift were expected to influence the evolution of a population: the research of Kerr and Wright (see chapter 3) and of Dobzhansky and Pavlovsky (chapter 17). Dobzhansky and Spassky reminded readers that the Dobzhansky-Pavlovsky experiment showed that large populations have smaller variations in genotype frequencies, as predicted, and reported another experiment confirming the prediction that "if the sizes of the foundation stocks of the populations are kept constant, the variations of the outcomes should be greater if these foundation stocks came from a source with higher genetic variability than from a genetically more uniform source" (Dobzhansky and Spassky 1962, 149). In the meantime, LaMotte (1959) had shown that the expected greater variation of smaller populations was also present in natural populations.

A different kind of prediction was made by Ford (1940a, 1945), based on Fisher's theory of balanced polymorphisms. He suggested that an association between blood groups and susceptibility to certain diseases would be found in humans. This was confirmed by Aird et al. (1953, 1954) and others in the 1950s (chapter 13). Many evolutionary biologists were aware of this success, yet like the experiments by Kerr and Wright and by Dobzhansky's group, it was rarely mentioned in the discussions with philosophers.

Popper versus Darwin

In 1945, the philosopher Karl Popper published the first of several critiques of the Darwinian theory of evolution. His article was primarily an attack on "historicism"—the assumption that history is determined by inescapable laws—in the social sciences. He stated that "the recent vogue of historicism," which he blamed for the evils of fascism and communism, "might be regarded as merely part of the vogue of evolutionism" (Popper 1945, 69). It was therefore necessary to discuss first the use of evolutionism in the biological sciences. He asserted that there could be no such thing as a scientific law of evolution, because the evolution of life on earth ... is a unique historical process.... We cannot hope to test a universal hypothesis or to find a natural law acceptable to science if we are confined to the observation of one unique process. Nor can the observation of one unique process help us to foresee future developments. (Popper 1945, 70)

Yet at the same time, he admitted, "I see in modern Darwinism the most successful explanation of the relevant facts" (Popper 1945, 69).

Popper's critique, published first in the journal *Economica* and then as a book, *The Poverty of Historicism*, did not attract much attention from evolutionists until it was printed in 1957, with a second edition in 1960. By that time, an English translation of his *Logik der Forschung* had appeared, containing a detailed exposition of his views on scientific method and on the use of falsifiability as a criterion to distinguish real science from pseudoscience (Popper 1959). Explanation of the facts, no matter how successful, was not enough: one must be willing to make testable predictions of previously unknown facts or events.

The American philosopher Michael Scriven then published an article in *Science* in which he argued, against Popper (and also against R. B. Braithwaite, as well as C. G. Hempel and P. Oppenheim), that evolutionary theory does indeed offer "satisfactory explanation of the past ... even when prediction of the future is impossible" (Scriven 1959, 477). Scriven's article elicited reactions from biologists and became part of the debate about the scientific legitimacy of evolutionary theory in the 1960s.

I see two serious problems with Scriven's argument. First, he defines a category of subjects, including "a great part of biology, psychology, anthropology, history, cosmogony, engineering, economics, and quantum physics," which he calls "irregular subjects"; in these subjects, "serious errors are known to arise in the application of the available regularities to individual cases," and he implies that "prediction is excluded" in these subjects (Scriven 1959, 477). Aside from the derogatory connotation of the word "irregular," his conception of prediction is impoverished. Quantum physics most certainly does make predictions that can be tested in the laboratory, and some of them have been confirmed with incredible accuracy. Apparently Scriven has in mind the fact that the present position or quantum state of an individual subatomic particle is indeterminate until it is measured, hence its future position or quantum state is unpredictable without a measurement, and the *presumed* fact that cosmogony deals only with unobservable and hence unknowable events in the distant past. (He had argued in 1954 that we will never be able to determine whether the age of the universe is finite or infinite.) Moreover, strictly speaking, *no* science, regular or irregular, can offer an accurate "prediction of the future," except in the sense that it can predict more or less accurately the result of a measurement we make in the present under controlled conditions. Guesses and extrapolations about future weather, earthquakes, and so forth may be useful in planning our lives but become increasingly unreliable after a few days or weeks. Yet meteorology and seismology are respectable sciences.

Thanks to the mathematical theories of Fisher, Haldane, and Wright and the availability of fast-breeding organisms like *Drosophila*, it was possible by the 1940s to test the predictions of evolutionary theory in precisely the same way that one could test the predictions of quantum mechanics and cosmogonical theories[185]—though philosophers, and even some biologists, seemed unaware of this fact.

Dobzhansky had already clearly distinguished between the historical approach to evolution (phylogeny) and the causal-mechanistic approach (population genetics).[186] As was subsequently pointed out by Ayala (1977), the second approach does use the hypothetico-deductive method; Popper and other philosophers who asserted that evolutionary theory is not falsifiable simply ignored the entire corpus of work for which Dobzhansky and his colleagues had become famous by 1959.[187]

Second, Scriven asserts that Darwin and others have attempted "to encapsulate the principles of evolution in the form of *universal* laws and base *predictions* on them" but have failed (Scriven 1959, 477). On the contrary, while I do not claim that biologists have established universal laws of evolution, I have shown that they have used the principles of natural selection to generate several predictions that have been tested and confirmed (or refuted, as in one case).[188]

A theory that governs the future behavior of a system, like Newtonian or quantum mechanics or general relativity, can also yield predictions about its present behavior.[189] This is a crucial fact often ignored by those who say that evolutionary theory cannot make testable predictions. It is ironic that Scriven lumps Darwinism with cosmology in the category of irregular theories that can explain but not predict. Within six years of the publication

185. Berry (1982, 57); see also Hull (1974) and Williams (1981).
186. Dobzhansky (1951, quotation in chapter 6); see also Prout (1995).
187. Ironically, a prominent member of a later generation of philosophers of biology accused Dobzhansky and Ayala of being *too* Popperian (Lloyd 1988, 10)!
188. Other early testable predictions were discussed by Lewis (1986); some later predictions from Fisher's theory are discussed by Leigh (1986).
189. Ruse (1969) and Prout (1995).

of Scriven's article, the prediction from the big bang theory (contrary to the steady state theory) that space is presently filled with microwave radiation at a few degrees above absolute temperature was spectacularly confirmed; this confirmation played some role in the decision of proponents of steady state to switch to big bang, although that decision was not greatly influenced by knowledge that the radiation had been predicted (Brush 1993a).

A related example, closer to the domain of biological evolution, is the theory of the origin of elements in stars. In order to explain the synthesis from hydrogen of elements heavier than helium, Fred Hoyle predicted a previously unobserved nuclear reaction, the tri-alpha reaction (fusion of three helium nuclei to form a carbon-12 nucleus), which was subsequently confirmed in a laboratory experiment. Even though the entire process of the evolution of elements from hydrogen takes too long to follow in the laboratory, almost every individual stage can be explained theoretically and tested experimentally (Brush 1996b, chapter 2.3).

Although many scientific theories do make successful predictions, that success is not necessarily a requirement for accepting those theories.[190] It does raise the barrier for anyone who wants to replace a successfully tested theory by an as yet untested one.

How Evolutionists Replied to Popper and Scriven

The biologists were too quick to concede Scriven's charge that evolutionary theory does not make testable predictions. I find it remarkable that Ernst Mayr, who had himself made a confirmed prediction from evolutionary theory (as he was pleased to remind the readers of his next book [Mayr 1963, 207]), agreed with Scriven:

> The theory of natural selection can describe and explain phenomena with considerable precision, but it cannot make reliable predictions. . . . Scriven has emphasized quite correctly that one of the most important contributions to philosophy made by the evolutionary theory is that it has demonstrated the independence of explanation and prediction.

But, he continued, we are happy if "our causal explanations [also] have predictive value" (Mayr 1961, 1504). Apparently what Mayr meant was that evolutionary theory can be scientific as long as it provides good explanations, even if it doesn't make testable predictions, but sometimes it does

---

190. See note 5.

make testable predictions. Later he distinguished *logical* prediction (defined as "conformance of individual observations with a theory or scientific law") from *temporal* prediction (inference from the present to the future). The latter (what physicists call "prediction in advance" and philosophers call "novel prediction") is "much more rarely possible in the biological sciences" (Mayr 1982, 57–59). But three years later, he argued that "in principle there is no difference between the physical and the biological sciences with respect to experiment and observation": neither makes "absolute" predictions (Mayr 1985, 50). Nevertheless, he agreed with Scriven that "the ability to predict is not a requirement for the validity of a biological theory" and stated: "The theory of natural selection can describe and explain phenomena with considerable precision, but it cannot make reliable predictions, except through such trivial and meaningless statements as, for instance, 'the fitter individuals will on the average leave more offspring'" (Mayr 1988, 19, 31).

Simpson also agreed with Scriven that a historical science like evolutionary biology or geology does not make predictions in the same way the physical sciences do, but argued that

> the testing of hypothetical generalizations or proposed explanations against a historical record has some of the aspects of predictive testing. . . . A conspicuous example has been the theory of orthogenesis, which . . . maintains that once an evolutionary trend begins it is inherently forced to continue. . . . That plainly has consequences that would be reflected in the fossil record. As a matter of observation, the theory is inconsistent with that record. . . . A crucial historical fact or event may be deduced from a theory and search may subsequently produce evidence for or against its actual prior occurrence. That has been called "prediction" . . . what is actually predicted is not the antecedent occurrence but the subsequent discovery. (Simpson 1964, 144–5, 147)

A few years later he wrote, "Claims that the hypothetico deductive method (e.g., Popper 1959) is the only one allowable in science are almost absurdly extreme, but obviously it is an allowable method. . . . Retrodictive interpretation and explanation are almost unique to the historical sciences."[191]

The most elaborate refutation of the philosophical critics in the period before 1970 was Michael Ghiselin's book, *The Triumph of the Darwinian Method*. Ghiselin accepted Popper's falsifiability criterion and insisted that Darwin's theory satisfies it:

191. Simpson (1970, 85, 90); see also Ayala (1968b) and Lewontin (1972).

It seems almost unreal that, among all theories of science, the Darwinian theory of natural selection has been singled out [e.g., by von Bertallanfy] as incapable of refutation. . . . A theory is refutable, hence scientific, if it is possible to given *even one* conceivable state of affairs incompatible with its truth. Such conditions were specified by Darwin himself, who observed that the existence of an organ in one species, solely "for" the benefit of another species, would be totally destructive of his theory. That such an adaptation has never been found is a most compelling argument for natural selection. (Ghiselin 1969, 63)

But Ghiselin did not discuss the more recent predictions by Dobzhansky, Ford, Mayr, and Wright.

Although several evolutionists believed they had already satisfied the demand for testable predictions and could do so in the future,[192] others were not satisfied that their theories were sufficiently predictive.[193] Rudwick (1964) argued that orthogenesis could be tested but that the synthetic theory could not (at least in the domain of paleontology). Waddington rejected Popper's criterion, arguing that "in practice," a hypothesis can't be disproved because it "can always be suitably amended to deal with the objections raised" (1969, 110). Conversely, it was possible, in biology as in other sciences, for a theory to make a confirmed prediction and yet still be wrong.[194]

Philosophers disagreed among themselves. E. Manser agreed with Popper that evolutionary theory is not predictive and therefore fails "to explain in the normal scientific sense" (1965, 31). Hugh Lehman asserted, *contra* Scriven, that "Darwin's theory *can* be put in such a form as to fit the [Hempel] covering-law model using the 'statistical-probabilistic pattern' rather than the deductive-nomological pattern. He continued by saying, "We also try to show that Darwin's principles can be used to make predic-

---

192. Bonner (1962); Dowdeswell (1963); Simpson (1964, 136); Simpson and Beck (1965); Wallace (1966); Baker and Allen (1967, 503); Carlson (1967, 305); Lewontin (1967); Sheppard (1967); Waddington (in Moorhead and Kaplan [1967, 98]); Dobzhansky (1968); Slobodkin (1968); and Smith (1969).

193. Lewontin (1965, 1968b, 1974); Medawar (in Moorhead and Kaplan [1967, "Remarks by the Chairman"]); Simpson (1964, 147); Waddington (in Moorhead and Kaplan [1967, 13, 98]). Note the references in this and the previous note to three biologists (Lewontin, Simpson, and Waddington) who took different positions on this issue in different places. There was also an "undercurrent of perplexed and subtle dissatisfaction with the concept" of natural selection, perhaps to be resolved by using the semantic "theory of signs" of C. W. Morris (Mason and Langenheim 1961, 158).

194. Dunn and Landauer (1934, 239) and Laudan (1981).

tions," but he does not give any examples of facts not already known (1966, 14–16). Dudley Shapere wrote that prediction is not impossible in biology, just difficult because of the number of interacting factors, *not* because of the "essentially historical character of biological explanation" (Shapere 1969, 14). A. G. N. Flew (1966) asserted that Darwinism *can* make testable predictions, but the examples he gave were not impressive and not supported by specific evidence: if bacilli develop resistance to one kind of antibiotic, then we may confidently expect that new antibiotics will eventually produce similar immunity.[195]

Thomas Anderson Goudge seemed to be satisfied that natural selection theory does make testable predictions; in particular, the theory of random genetic drift predicts that a population of only one hundred animals will probably become extinct, while a population of one hundred thousand sub-divided into one thousand subpopulations of one hundred each will probably enjoy "sustained, relatively rapid and progressive evolution" (Goudge 1961, 126).

Michael Ruse pointed to the return of light-colored moths to "those areas that have introduced smokeless zones" as a fact that was "clearly expected" (and thus could have been predicted) by the synthetic theory (Ruse 1969, 339); the case of wingless insects is also evidence for natural selection, he noted. Ruse later criticized Popper's views on Darwinism (Ruse 1977).

Mary B. Williams, in the first of a series of papers on this subject, proposed an axiomatization of Darwinian theory, which "provides the foundation for a truly predictive (and therefore testable) theory of evolution" (Williams 1970, 370). David Hull (1974) cast doubt on the basic premise that biology is qualitatively different from physics with respect to predictions.

But with the exception of Ruse's brief remarks about wingless insects, philosophers did not discuss the confirmed predictions of natural selection mentioned in this monograph, at least in the literature of the 1960s that I have been able to examine.[196]

In my opinion, the most satisfactory response to Popper's criticism of Darwinism was the book by philosopher Stephen Toulmin, who does not mention it at all. Toulmin asserts that it is futile to seek a simple criterion such as predictive success to determine whether a theory or activity is "scientific."

195. See comments by Barker (1969).

196. References for philosophical discussions of predictiveness in biology can be retrieved from sources such as the *Philosopher's Index* and the books by Hull (2001); Hull and Ruse (1998); Lloyd (1988); and Sober (1984a, 1984b, 1993). For further discussion of the testability of evolutionary theory, see Williams (1973, 1982, 1985) and Zachos (2002).

The critical questions which a philosopher brings to science must be co-ordinated with the factual studies of history. We must be prepared to scrutinize in detail a representative selection of classic theories, analysing critically the merits for which they came to be accepted, and the standards by which they established their claims to superiority. (Toulmin 1961, 16)

Forecasting . . . is . . . an application of science, rather than the kernel of science itself . . . a novel and successful theory may lead to no increase in our forecasting skill; while, alternatively, a successful forecasting technique may remain for centuries without any scientific basis. (Toulmin 1961, 36)

Thus, Darwin's theory cannot "foretell the coming-into-existence of creatures of a novel species." Yet Darwinism did forecast "short-term, small-scale events":

When Australians used myxomatosis [a viral disease] to control the rabbit population, it was forecast on the best Darwinian principles that a new strain of rabbits would become dominant, whose constitutions were more resistant to the disease. . . . The correctness of this prediction has helped to confirm the merits of the Darwinian theory. And the same thing has happened in other small-scale cases involving melanism in moths, the reactions of infective micro-organisms to anti-biotics, and so on. (Toulmin 1961, 25)

Mathematicians and Physicists
Challenge Natural Selection

Dissatisfaction with the supposedly nonpredictive nature of evolutionary theory and the belief that biologists might be able to learn something useful from physicists and mathematicians led some of them to organize a symposium on "Mathematical Challenges to the Neo-Darwinian Interpretation of Evolution," held at the Wistar Institute of Anatomy and Biology in Philadelphia, April 25–26, 1966.[197] As it turned out, the symposium

197. Martin Kaplan, coeditor of the proceedings, wrote in its preface that "the seed was sown in Geneva in the summer of 1965 . . . when [Hilary] Koprowski [director of the Wistar Institute] was confronted by a rather weird discussion between four mathematicians . . . on mathematical doubts concerning the Darwinian theory of evolution" (Moorhead and Kaplan 1967). Peter Medawar, chairman of the symposium, said "the immediate cause of this conference is a pretty widespread sense of dissatisfaction about . . . the so-called neo-Darwinian [t]heory." Apart

. . . .

was not devoted primarily to the topic of its title. The mathematicians and physicists had apparently never looked at the canonical works of Fisher, Haldane, and Wright (as Ernst Mayr pointed out in one of the discussions) but instead tried, rather clumsily, to reinvent the theoretical wheel. What was more interesting (at least to me) was what Ernst Mayr called "Evolutionary Challenges to the Mathematical Interpretation[s] of Evolution," as those interpretations were presented by Murray Eden, professor of electrical engineering at the Massachusetts Institute of Technology; Stanislaw Ulam, mathematician at Los Alamos and coinventor of the hydrogen bomb; William Bossert, computer expert at Harvard; and Marcel Schützenberger, professor of mathematics at the University of Paris. Among the more active participants in the discussion was V. F. Weisskopf, a well-known nuclear physicist at the Massachusetts Institute of Technology and one of the original instigators of the meeting.

The biologists at the symposium—especially Mayr and Waddington—argued that the models proposed by the mathematicians were grossly oversimplified, leaving out such essential factors as mutation rates and population size. Ulam's model, for example, led to an intractable equation that did not apply to sexually reproducing organisms, but by making a very rough approximation, he was able to arrive qualitatively at the conclusion that including sexual reproduction would speed up evolution. Waddington asked, "How [does] this [differ] from the standard biological theory that sexual reproduction speeds up evolution enormously?" Ulam replied, "Obviously common sense tells us [this] immediately; the whole point of pursuing a mathematical model is to get some quantitative estimates." Waddington and geneticist Niels Barricelli then remarked that this was exactly what Fisher, Haldane, and Wright had done—much more accurately and with simpler mathematics—thirty-five years earlier. Mayr and other biologists repeated that they had already arrived at the same conclusions *without* knowing any mathematics and without using (as Ulam did) a model that failed to make biological sense (Moorhead and Kaplan 1967, 21–33).

I suspect the net result of the symposium was to convince the biologists that they already had all the math they needed from Fisher, Haldane, and

---

from religious objections, such as "the kind of pious bunk written by Teilhard de Chardin," the "philosophical and methodological objections ... have been very well voiced by Professor Karl Popper—that the current neo-Darwinian [t]heory has the methodological defect of explaining too much. It is too difficult to imagine or envisage an evolutionary episode which could *not* be explained by the formulae of neo-Darwinism." In addition, "there are objections made by fellow scientists who feel that, in the current theory, something is missing.... These objections ... are very widely held among biologists generally" (Medawar 1967, p. xi).

Wright, and they did not need to feel inferior because they couldn't derive the equations themselves.

The outcome of Popper's critique of evolutionary theory, mentioned by several of the mathematicians and physicists at the symposium, was less fortunate. Popper apparently never learned about the successful predictions that evolutionists had actually made; he repeated his claim that Darwinian theory is not scientific, and even though he eventually retracted it (Popper 1978), the damage had been done: another weapon had been handed to the creationists in their battle against evolution.[198]

198. In 1981 the Arkansas state legislature passed a law (Act 590) demanding equal time for creationism and evolution, although it was later ruled unconstitutional. The law included the phrase, "evolution cannot be experimentally observed, fully verified, or logically falsified" (chapter 7); the last word clearly indicates that the creationist writers of the law had exploited Popper's critique (La Follette 1983). A similar bill proposed in (but not passed by) the Maryland General Assembly used a similar phrase.

22

....

Context and Conclusions

In the first part of this chapter, I discuss some speculations about possible cultural and philosophical influences on evolutionary theory; then I return to a survey of the scientific reasons for accepting the natural selection hypothesis.

Jonathan Harwood has called attention to what he calls the "metaphysical foundations of the evolutionary synthesis," by which he means attitudes toward holism, reductionism, materialism, and other concepts that may have encouraged or discouraged the formation and reception of the synthesis at specific times and places. He suggests that the ideology of *Bildung* (cultivation) and the calls for holistic anti-reductionist synthesis in interwar Germany "might have prompted Dobzhansky, Mayr, Simpson, et al. to adopt a synthetic framework in tackling the problems of evolutionary theory" (Harwood 1994, 5). Intellectuals influenced by this tradition tend to have broad cultural as well as scientific interests and are attracted to the "romantic tradition in biology," as exemplified by Goethe. Harwood also suggests that there may be some similarity between these influences and the ones that acted on German geneticists, as he described in an earlier work (Harwood 1993), and on physicists in the first part of the twentieth century, as discussed by Paul Forman (1971). Similarly, Colin Divall (1992) argues that Julian Huxley, a proponent of the "synthesis" label, was sympathetic to the philosophical idealism of A. N. Whitehead, Arthur Eddington, and James Jeans and to the "emergent evolution" concept of C. Lloyd Morgan, as a reaction against (Herbert) Spencerian mechanistic materialism. Harwood used the phrase "styles of thought" to characterize these metaphysical assumptions, and that phrase has become popular among historians of science in the last two decades.[199]

199. An entire issue of the journal *Science in Context* was devoted to "Style in Science" (Daston and Otte 1991). For applications to evolutionary biology, see Gerson (1998) and Bowler (2004). Richard Lewontin has also endorsed the usefulness of the concept of "epistemic style," as presented by Jane Maienschein (Lewontin 1995, 96; Maienschein 1991). A related term is

....

The word "synthesis" is misleading here. In the 1930s, an evolutionary theory that synthesized selection with Lamarckism, drift, mutations, and orthogenesis might have suited the spirit of the time. But it was clear by the 1950s that the leaders of the neo-Darwinian synthesis really wanted to synthesize evidence from different fields of biology to support a unified evolutionary theory based primarily on natural selection. When they concluded that most of this evidence could be explained without invoking drift, that large mutations were almost always fatal while there was an ample supply of small mutations to drive evolution, that Lamarckism was not supported by reliable evidence, and that orthogenesis was not a respectable scientific concept, they ended up with a hardened synthesis that could be considered reductionist and materialist. As Provine pointed out, it was not really a synthesis but was actually a constriction. But such a theory was quite compatible with the style of thought popular in post–World War II science and culture: not romantic but realist, sometimes known as "midcentury modernism." It was the age of physics based on elementary particles, quantum chemistry based on molecular orbitals, geology based on plate tectonics, molecular biology based on DNA, and analogies between brains and computers.[200]

For the present account, it seems most pertinent that the idea of randomness (the basis of the genetic drift concept) was strongly associated with a holistic, anti-reductionist, anti-mechanist metaphysics, which arose in German-speaking countries after World War I, especially in the Weimar Republic. It was, according to Paul Forman (1971), a response to cultural criticism of scientists and mathematicians who claimed to know everything. It was also associated with the radical anti-realist "Copenhagen interpretation" of quantum mechanics (Brush 1980). The same movement, which may be regarded as a revival of the romantic and neo-romantic movements of the nineteenth century (Brush 1978), has been shown by Anne Harrington (1996) and other historians of science to have had a strong influence on German biology in the 1930s. Garland Allen (1983) portrays I. Michael Lerner, who was born in Manchuria and worked in Canada before settling in theUnited States, as an example of the influence of "holistic materialism" in the form of "genetic homeostasis." But there is no reason to think it had any influence on Sewall Wright in the United States; in fact, Wright's biographer argues that Wright's own philosophical views had no discern-

---

"thema," introduced by Gerald Holton (1988) to identify concepts like "force" or "atom" or "evolution," whose meanings go beyond observation and mathematical analysis. On the relation between epistemic styles and themata, see Holton and Brush (2001, chapter 33).

200. Brush (1980).

ible effect on his scientific work (Provine 1986, 96–97). Wright did not invoke Heisenberg's indeterminacy principle, although Muller (1929, 498) and Fisher (1934) recognized a possible connection between indeterminism at the atomic level and indeterminism in evolution. But I would argue that one should look for connections between the holistic anti-reductionist viewpoint in biology and the eager acceptance of genetic drift—not as part of a catalyst for natural selection, as Wright himself proposed, but rather as an all-purpose (and somewhat vacuous) explanation for nonadaptive characters and species differences. At the same time, one should allow the possibility that oscillations in the popularity of drift are also driven by new empirical findings (Futuyma 1979, 272–6).

Similarly, as Simpson pointed out, the idealist philosophy of German biologists often led them to support a saltationist (macromutation) view of evolution. But he seems surprised that the same biologists (especially Rensch) could also support an empirical approach:

> To me ... it came as a really unpleasant shock to find Rensch avowing himself an idealist after a long discussion of evolution that could be endorsed, word for word, by the most rigid materialist ... [but] almost all of us [Americans?] share with Rensch a basically inductive approach to evolution. (Simpson 1949a, 184)

From my perspective, it is not at all surprising to find empiricism combined with idealism in science, since that is just what happened in late nineteenth-century neo-romanticism; nor is it surprising to find that empiricism can facilitate a transition from romanticism to realism (in this case, the transformation of evolutionary theory from the 1930s to the 1950s), since that is just what happened in the establishment of the modern energy concept earlier in the nineteenth century (Brush 1978). Indeed, the unification of biology by the evolutionary synthesis is in some ways similar to the unification of physics by the law of conservation and transformation of energy.

Other philosophical classification schemes may be useful here. For example, according to John Greene, "The champions of the modern synthesis ... occupied a polemical position midway between the reductionists and the vitalist-finalists." They used the language of purpose but denied its existence.[201]

Fisher's theory was of course also based on the assumption that genes are shuffled randomly during reproduction, but in a way that makes the

---

201. Greene (1986, 220); see also Greene (1999).

outcome much more predictable. Just as in the kinetic theory of gases, his model for population genetics, there are always so many particles that fluctuations are relatively small, and most properties of the system appear to be determined by a few parameters. The kinetic theory is a legacy of the realist-atomist period of the mid-nineteenth century and was revived (by Fisher's mentor Jeans) in the early twentieth century, just as Mendel's realist-atomist theory of heredity was revived. One might guess that the evolutionary theories of Fisher and Haldane reflected the metaphysical assumptions of this "neorealist period" (before World War I) and therefore were not very popular with most biologists until the 1940s and 1950s, when biology returned to the mechanistic-reductionist approach, perhaps best exemplified by the Watson-Crick double helix model of DNA.[202]

As Richard Lewontin wrote in 1968,

> the twenty years since World War II have seen a vindication in biology of our faith in the Cartesian method as a way of doing science . . . an analytic technique . . . the belief that by breaking systems down into their component parts, by simplifying them or using simpler organisms, one can learn about more complex systems. . . . It is *not* the case that molecular biology is Cartesian and analytic while population biology is holistic. Population biology is properly analytic and operates . . . by the process of simplification, analysis, and resynthesis. (Lewontin 1968a, 1–2)

But of course no movement—holism or reductionism, romanticism or realism—ever gains unanimous support at any time; there is always an opposition party to be denounced:

> It is unhappily true that there are population biologists who reject the analytic method and insist that the problems of ecology and evolution are so complex that they cannot be treated except by holistic statements. The influence of these people has held up progress in population biology for many years and, in addition, has tended to degrade population biology as a science. They are the stamp collectors of biology who, because they themselves are unable to analyze the complex problems of ecology and evolution, try to convince the rest of us that nothing but "objective description" of nature is possible. (Lewontin 1968a, 2)

---

202. See Depew and Weber (1985b, 228–9 [and elsewhere]). The idea of a "genetic code" based on the structure of a large molecule was in part inspired by the book *What is Life?* by Erwin Schrödinger (1944), the physicist who advocated (unsuccessfully) a realist interpretation of his wave equation.

And if they wait long enough, the opponents can catch the next swing of the pendulum back in their direction.

The pendulum may swing faster in other fields. "Selection" seems to have been a popular thema in the behavioral and social sciences after World War II, according to publications by the psychologist Donald T. Campbell in that period. Speaking at a 1961 conference on social change and evolutionary theory, he noted that there was currently a revival of interest in this topic, after decades in which social scientists rejected evolutionary theory because of its association with reactionary political views such as social Darwinism. The revival was not credited to any newly discovered scientific evidence for evolutionary theories but simply to the tendency of social scientists to reject the status quo, whatever it may be; bored with the anti-evolution stance of the previous generation, they now accepted evolutionism.[203] Other psychologists used natural selection as a model for trial and error learning (and, not mentioned by Campbell, for the modification of behavior by "operant conditioning").

According to Hamilton Cravens (1988), the de Vries mutation theory was associated with extreme hereditarianism in the nature versus nurture controversy of the early twentieth century; its replacement by natural selection in the 1930s and 1940s was accompanied by, and supported, a shift toward the recognition of environmental influences in the behavioral and social sciences. Fisher's advocacy of eugenics reminds us that such correlations are seldom perfect.

Philosophers used the selection analogy to describe the history of science and technology. Stephen Toulmin argued that evolutionary biology provides the best analogy for the development of scientific ideas: assessing the merit of a theory depends on "seeing in how many ways a novel scientific idea may . . . be better adapted than its predecessors or rivals" (Toulmin 1961, 17; see also Toulmin 1966).

Karl Popper, despite his criticism of Darwinism as a nonfalsifiable theory, considered it a plausible mechanism for the growth of knowledge (Popper 1972). In fact, Popper, in a lecture delivered in 1961 but not published until 1972, described his method of "conjectures and refutations" as a kind of natural selection. A hypothesis that survives tests and criticism better than its competitors "may, temporarily, be accepted as part of current scientific teaching" (Popper 1972, 261).

As for Darwin's theory of *biological* evolution, Popper made an interest-

---

203. Campbell (1960, 1965, 1970); see the essays by David Hull and others in Heyes and Hull (2001) and Popper (1972).

ing statement (probably not original but worth repeating): "Darwin's theory of natural selection showed that it is *in principle possible to reduce teleology to causation by explaining, in purely physical terms, the existence of design and purpose in the world.*" This suggests a connection with the fact, well known to theoretical physicists since the seventeenth century, that a large number of natural phenomena can be described in two ways that are mathematically equivalent but philosophically quite different: teleological variational principles and mechanistic causal equations. As the great Swiss mathematician Leonhard Euler wrote in 1744, "All effects of the universe can be explained equally happily from final causes [teleology] by the method of maxima and minima and from the effective causes themselves" (Truesdell 1960, 200–1). Two hundred years later, Princeton biologist Edwin G. Conklin invoked the authority of Einstein, Bohr, and Planck to support his view that both natural selection and purpose are factors of evolution: "Mechanism and teleology are complementary views of nature, neither excluding the other" (Conklin 1944, 132).

The earliest and simplest variational principle was Pierre de Fermat's principle in optics, which asserted that if a ray of light travels from point A in one medium to point B in another in which its speed is different, it will follow the path that makes the total time as short as possible. This seems to imply that the path of the ray is *designed* to achieve a future goal. Yet it gives the correct answer (Snell's law of refraction), previously derived by Descartes from a mechanistic particle model and later derived by Huygens from a mechanistic wave model. Descartes had to assume that light moves faster in a denser medium, which is hard to believe, whereas Huygens could make the more reasonable opposite assumption. So, Huygens succeeded in reducing Fermat's teleological explanation to physical causation; in a mathematical sense, one could say that Darwin is to Paley (the advocate of intelligent design) as Huygens is to Fermat. Of course no sensible physicist today believes that the ray of light is constructed by an intelligent designer to get from A to B as fast as possible, and no reasonable biologist today believes that organisms have been constructed by an intelligent designer to have the properties they do. Darwin's theory, like Huygens's, shows how the *appearance* of design can be *simulated* by a causal mechanistic process (cf. Sober 1996).

Reasons for Accepting the Natural Selection Hypothesis

Enough of metaphysics; let us now turn to table 5 (p. 135), where I have tabulated the most popular reasons for accepting the natural selection hypothesis—that is, the empirical evidence most often mentioned in Anglo-American books and technical reviews on evolution by biologists (now including "Founders and Leaders") in the period 1951–1970.

Notice that while Kettlewell's peppered moth is the most popular overall, it did not play a very significant role in these books until after 1955.[204] So if one wants to understand why most biologists had accepted the primacy of natural selection by 1960, one has to start with the more traditional illustrations: various kinds of protective coloration, including industrial melanism, as observed before Kettlewell's research. The recent controversy about the validity of Kettlewell's work is therefore largely irrelevant to the scientific status of natural selection in the 1960s.[205] Similarly, the new, more rigorous research on mimicry by Jane Van Zandt Brower and others, published in the late 1950s and afterward, came too late to have much effect in *changing* the views of evolutionists (Brouwer 1958, 1960). The same is true for Cain and Sheppard's work on snails, the connection between sickle cell and malaria, and the connection between blood groups and diseases. Instead, we have two rather different kinds of evidence playing an important role in addition to protective coloration: on one hand, the development of insecticide-resistant pests (going back to the scale insects in the citrus groves of California in the beginning of the twentieth century and the similar development of resistance to DDT and antibiotics in the 1940s and 1950s);[206] and on the other hand, the experiments of Dobzhansky and others on cyclic variations of chromosome inversions in *Drosophila* and *Adalia punctata*. Both show that natural selection acts much more quickly than had previously been believed—a few years in the first case, a few weeks or months in the second. The corresponding data for textbooks, popular articles, and books are shown in table 4 (p. 134). The most striking difference is the greater importance given to protective coloration and wingless insects, and the much smaller weight for chromosome inversions, snails, sickle cell/malaria, and blood groups.

By the 1970s, it was not enough to show that natural selection was more important than other evolutionary processes; the goal posts had been moved, thanks in part to the success of the NSH itself. W. H. Dowdeswell complained that there was so far very little "precise evidence for the action of selective agents" (Dowdeswell 1971, 369). Fifteen years later, John Endler compiled a list of more than one hundred empirical demonstra-

204. See chapter 19; of the seven references to industrial melanism in the period 1951–1955, only one (Ford 1955) specifically mentioned Kettlewell's work.

205. Hooper (2002) and Rudge (2000b, 2003, 2005).

206. Simon (2003 [and works cited therein]) and Creager (2007).

tions of natural selection in the wild; most of them were published in the 1960s or later and thus had little influence on the earlier acceptance of the NSH (Endler 1986 126–53). Even the more recent ones did not, he argued, generally answer the important quantitative questions about selection.

Some of Dobzhansky's early experiments were designed to test genetic drift, and contrary to his expectations, they showed that the effect of drift was negligible compared to that of natural selection; as a prominent leader of the evolutionary synthesis, his conversion to the natural selection hypothesis undoubtedly influenced many other biologists. The experiments in the 1950s by Kerr and Wright and by Dobzhansky and Pavlovsky showed that the effects of genetic drift *in combination with natural selection*, predicted by the Fisher-Haldane-Wright theory, could be detected by *specially designed laboratory experiments* with small populations. But there was no credible evidence that genetic drift by itself played a significant role in evolutionary processes in nature.

The fact that old evidence, going back to the period before 1930, was so frequently invoked in support of the new version of natural selection theory leads us to ask: did the new evidence really play a major role in persuading evolutionists to accept natural selection? I suggest two answers to that question.

First, it seems clear to me that the mathematical labors of Fisher, Haldane, and Wright were essential to the success of the new natural selection hypothesis and indeed to the success of the evolutionary synthesis itself. One could support this view by quoting Dobzhansky, Huxley, Simpson, and many others.[207] But perhaps the best authority is Ernst Mayr, because he is notorious for his attempts (along with those of C. H. Waddington) in the 1950s to minimize the contributions of the mathematicians, relative to those of the field and laboratory biologists, and for his ridicule of the "beanbag genetics" of Fisher, Haldane, and (unfairly) Wright.[208] Mayr was the senior author of a zoology textbook in which one finds, near the beginning, the flat statement:

Fisher's (1930) demonstration that even a very small selective advantage of a new gene or gene combination would cause in due time a genetic transformation of populations was a tremendously important contribution. (Mayr et al. 1953, 12)

207. Dobzhansky (1955, 14); Huxley (1942, 123); Provine (1978); Sheppard (1954, 201–2); Simpson (1949b, 266); and Watson (1951, 192).
208. Mayr (1959) and Provine (1986, 480–4).

. . . .

At the end of our period, Mayr summed up the situation as follows:

> The modern attitude toward natural selection has two roots. One is a mathematical analysis (R. A. Fisher, J. B. S. Haldane, Sewall Wright, and others), demonstrating conclusively that even very minute selective advantages eventually lead to an accumulation in the population of the genes responsible for these advantages. The other root is the overwhelming mass of material gathered by naturalists on the effect of the environment. This evidence was given a largely Lamarckian interpretation in the days when mutations were claimed to be saltational and cataclysmic. When small mutations were discovered, and when it was realized that all variation had ultimately a mutational origin, this evidence became a powerful source of documentation for the selectionist viewpoint. (Mayr 1970, 108)

So, the phenomena of protective coloration, mimicry, and development of resistance to pesticides and antibiotics *became* evidence for natural selection, at least in part because of the theoretical work of Fisher, Haldane, and Wright. Fifteen years later, industrial melanism and the development of resistance were still considered the best empirical evidence for natural selection to present in an introductory course (Moore 1984, 515–6).

This evidence has also had an impact on the ongoing controversy over the teaching of evolution and creationism in American public schools. Eugenie C. Scott, a leader of the pro-evolutionists in this controversy, wrote that "the evidence for the operation of natural selection is so overwhelming that both IDCs [Intelligent Design Creationists] and YECs [Young Earth Creationists] now accept that it is responsible for such phenomena as pesticide resistance in insects or antibiotic resistance in bacteria" (Scott 2004, introduction). When pressed to account for these phenomena, whose importance to medicine and agriculture is obvious, creationists like Henry Morris make an arbitrary distinction between microevolution (which undeniably does occur) and macroevolution (which, they assert, does not).[209]

The second reason is mentioned in Mayr's 1970 summation quoted previously: H. J. Muller's discovery of radiation-induced mutations provided an independent confirmation of Fisher's geometrical argument that only very *small* mutations can contribute to evolution. Large mutations may occur but are likely to produce death or sterility. Those biologists who rejected Lamarckism and orthogenesis for other reasons were now forced to reject also the de Vries version of neo-Darwinism and were left with the

---

209. Morris (1974) and many articles in *Acts & Facts*, the monthly newsletter of the Institute for Creation Research. See Numbers (2004) for the acceptance of microevolution by creationists.

Fisher-Haldane-Wright version of natural selection as the only plausible alternative.

Third, Dobzhansky's research on *D. pseudoobscura* forged an important link between laboratory genetics and field observations, allowing evolution to become an experimental discipline in which hypotheses about natural selection and drift could be tested.

The fourth answer, which helps to explain the success of the natural selection hypothesis, is what might be called the "Lakatos effect." Imre Lakatos was one of the first philosophers of science to point out that scientists do not simply weigh the evidence on both sides of an issue; they also compare the track records of competing theories (or what Lakatos called "research programmes"—series of theories). If theory A is found to successfully predict or give a better explanation of a phenomenon previously thought to be better explained by theory B, it thereby becomes more "progressive," while theory B becomes more "degenerating." If scientists are persuaded that all of the *recent* research has transformed evidence for B into evidence for A, they are likely to switch their allegiance to A, *even if most of the evidence still favors B*, because they anticipate that A will eventually win out, and this can become a self-fulfilling prophecy. In fact, during the 1950s and 1960s, several supporters of natural selection (theory A) claimed that this was the situation: since many characters previously thought to be nonadaptive and therefore determined by genetic drift (theory B) had been shown by more detailed research to be adaptive, while there were no examples of the opposite; *therefore*, the default assumption should be that all characters are adaptive (or genetically linked to other characters that are adaptive) in the absence of definite proof that they are nonadaptive. To the extent that evolutionists accepted this claim, the natural selection hypothesis was considered progressive, while the genetic drift theory was considered degenerating—even though there might still be a large number of cases of apparently nonadaptive characters for which no selective basis had yet been found.[210] Similarly, H. J. Muller argued that when we see phenom-

---

210. Beatty (1987b, 57); Cain (1951a; 1951b; 1964: 37); Darlington (1963, 61–62); Dobzhansky (1962, 280; 1970, 262); Dunn (1959, 103); Ford (1964, 35; 1971, 39–41); Lack (1961, preface); Mayr (1970, 120–4); Sheppard (1958, 123; 1967, 110); and Stebbins (1959a, 50; 1966, 74–76; 1970, 184–5). Another version of this effect is described by John R. G. Turner: "Repeated observations [by the Oxford School] of rather strong selective effects, too strong to allow significant drift, or producing clear adaptive differences between local populations, left an onus of proof on the supporters of Wright's theory" (1987, 342).

Philosophers are still debating about how to define drift and whether it can be clearly distinguished from natural selection as a separate force or factor in evolution: Sober (1984); Millstein (2002, 2005); Walsh et al. (2002); Sober (1984a); Stephens (2004); Brandon (2005); Pfeifer (2005); and Plutynski (2007b).

ena that *look* orthogenetic, we should not postulate "mysterious processes" like orthogenesis just because we have not *yet* explained them by natural selection; the success of natural selection in explaining other phenomena makes it the preferred default hypothesis (Muller 1949, 440).

Of course the Lakatos effect can help either side; it gives an advantage to the side whose evidence is more *recent*. For example, in the late nineteenth-century debate about the atomic structure of matter, the atomists had shown around 1870 that evidence from several different phenomena led to similar estimates for the size of an atom and for Avogadro's or Loschmidt's number. Despite this, the anti-atomist thermodynamics program seemed to be more progressive at the end of the nineteenth century, at least by Lakatosian standards (Clark 1976). Then, thanks to the development of new theories of Brownian movement by Albert Einstein and Marion von Smoluchowski, Jean Perrin was able to obtain new estimates of the size of the atom and of Avogadro's number, which agreed fairly well with the old estimates; on the basis of this agreement and other experiments, he was able to establish once and for all the atomic nature of matter.[211] The new evidence was not really much better than the old; it was just newer.

The alternatives to the natural selection hypothesis included, in addition to genetic drift, Lamarckism and orthogenesis. Throughout the period 1930–1970, Lamarckism was almost universally dismissed (except in France and in German paleontology) as unproven and implausible. The only advocate who was taken seriously by biologists was the psychologist William McDougall, whose experiments on the training of rats were repeated and finally discredited in a two decades-long study by a group at the University of Melbourne (Agar et al. 1954). Orthogenesis had been popular in the 1930s but was strongly attacked by leaders of the evolutionary synthesis as a vestige of unscientific metaphysical or vitalist thinking; Simpson persuaded some, but not all, paleontologists that their observations could be better explained by natural selection.[212]

A fourth alternative, the view that macromutations are essential for speciation, continued to attract a small minority of biologists. The only major

211. Brush (1968, 1976) and Nye (1972).

212. Another theory sometimes discussed was "organic selection," also known as "the Baldwin effect," "canalization," or "genetic assimilation." I have not discussed it in the text because it rarely appeared in textbooks. The literature on this theory is rather confusing, but it appears to be an attempt to explain by natural selection certain adaptations that seem at first glance to be Lamarckian in nature or to involve some behavior by the organism that can guide its own evolution (Waddington 1961). Waddington's experiments in the 1950s on the development of the "veinless" characteristic in *Drosophila* subjected to high temperatures were highly regarded (Huxley 1963; Merrell 1962; Moody 1962), but his attempt to explain the results by genetic assimilation was

advocate of this view was Richard Goldschmidt, and like Wright's genetic drift, he regarded macromutations as a supplement rather than a replacement for natural selection. Like drift and orthogenesis, macromutation was an idea congenial to the holistic German biology of the 1930s. As Goldschmidt himself recognized, "All theories of evolution tend to reflect the scientific trends of the time" (Goldschmidt 1940, 397), and the trend of the 1940s and 1950s favored the natural selection hypothesis. But Goldschmidt was too important a biologist to be ignored, so his theory had to be given the courtesy of serious consideration—followed by definitive rejection. Macromutations were not important in evolution because: (1) they are always lethal (Fisher 1930a; Muller 1938); (2) there was no solid empirical evidence that speciation worked that way; (3) there was no reason to make a qualitative distinction between micro- and macromutation, because every step in between had been observed and seemed to follow the same rules as micromutations; (4) there was no need to postulate macromutations, because micromutations could account for all evolutionary processes, as soon as one recognized the tremendous power of selection; and (5) research in molecular biology following the establishment of the double helix model of DNA was incompatible with the existence of macromutations that could produce new species.

As previously noted, at least three predictions based on natural selection were confirmed between 1930 and 1970: Darwin's prediction that flying organisms on a windy island would evolve a wingless form; Mayr's prediction that *Drosophila* chromosome inversions would prove to be adaptive (or if you prefer, Dobzhansky's refutation of his own expectation of the opposite result); and Ford's prediction that certain blood groups would be associated with disease. Aside from Mayr and the Ford school, evolutionists gave these successful predictions no extra credit for novelty and did not even mention them in the debate with philosophers about whether evolutionary theory is falsifiable. Nor did they mention the confirmations of evolutionary theory's predictions about the effects of genetic drift by Kerr and Wright, by Dobzhansky and Pavlovsky, and by Lamotte. Bruce Wallace seems to have been the only exception, but his statement that the confirmation of a prediction is "satisfying" referred not to natural selection but to Dobzhansky's prediction that certain previously unobserved chromosome arrangements would be discovered because they are needed as intermediate

---

controversial. The use of the concept was strongly criticized by Simpson (1953b), Mayr (1963), and Williams (1966); Moody was persuaded by these criticisms to withdraw his support in the third edition of his textbook (1970). See also Schmalhausen (1949); Richards (1987); B. Hall (2001); and Weber and Depew (2003).

stages in the evolution from one known arrangement to another. Moreover, this statement about the value of a prediction was made in a popular work, not one intended for other biologists.

If one wanted to make the case that evolutionists were converted to the natural selection hypothesis by following the scientific method, the best argument would have to be based on the outcome of the second prediction. Dobzhansky did, in effect, make a prediction; he did compare it with observations, and then he tested the conclusion by controlling the variables in a laboratory experiment, which indeed showed that the proposed selective agent (seasonal temperature changes) did produce the predicted result.[213] Moreover, Dobzhansky himself stated that the result of this research led him to adopt the natural selection hypothesis (which he had previously doubted), and the research was mentioned in this connection by several authors of books on evolution. On the other hand, this was originally a "contraprediction"—a test of the prediction that, contrary to the natural selection hypothesis, genetic drift does have a significant effect on the evolution of gene frequencies. When the contraprediction was refuted by field observations, he then invented a possible selective agent and showed that it was correlated with changes in gene frequencies. This sequence of events seems consistent with Popper's idea that the predicted result should ideally be one that would *not* be expected in the absence of the hypothesis. Yet only a minority of authors of books on evolution even mentioned this test (table 2); and as far as I know, none of them asserted that the result was persuasive *because* it was a successful *novel* prediction. Moreover, when philosophers and a few biologists criticized evolutionary theory for not being testable, no evolutionist seems to have brought forward this example in response.[214] I can conclude only that while biologists do make testable predictions and sometimes confirm them, they do not (despite the rhetoric of textbook writers in their introductory chapters on scientific method) really believe

213. In retrospect one could argue that strictly speaking, this was not a crucial test and that the laboratory experiment was not really a replication of the field results (Lewontin 1995), but it did seem to have an impact on the views of Dobzhansky himself and others.

214. In a recent "episodic history" of philosophical issues in biology, Marjorie Grene and David Depew (2004, 249) mention the question, "Is evolutionary theory predictive? If it is not, how can it be explanatory?" They state that Popper called natural selection nonfalsifiable, partly because he misunderstood the new Darwinism; "Overstressing the chance element, he thought that 'each small step [leading to an adaptation] is the result of a purely accidental mutation'" (Popper 1972, 269–70 [as quoted by Grene and Depew 2004, 262]). But Grene and Depew fail to mention any of the confirmed predictions of evolutionary theory, and the reader may be left with the impression that the view of Mayr and Lewontin—that biology is not a law-based science and does not need to be predictive—is generally accepted (Grene and Depew 2004, 265–6; Mayr 1985, 1988; Lewontin 1991).

that predictiveness is the most important criterion for judging a new theory. On the other hand, they expect that an established theory should be able to make predictions and that if one of those predictions is falsified, one need not necessarily abandon the theory itself but perhaps make minor adjustments to accommodate the data. In this respect, they are very much like physical scientists. The difference is that, at least during this period, their predictions were more qualitative than quantitative.

ACKNOWLEDGMENTS

....

This book is greatly indebted to the pathbreaking publications of Will Provine and to his extensive critique of an earlier draft; I regret that I have not been able to master the subject as thoroughly as he has. I got a late start: high school biology circa 1948 was incredibly boring, a result of the lingering effects of the Scopes Trial (Grabiner and Miller 1974), so I avoided taking any biology in college.[215] I can testify to the truth of Dobzhansky's (1973) maxim, "Nothing in biology makes sense except in the light of evolution." Lindley Darden also provided a very helpful and detailed critique, although I have not quite been able to bring my exposition up to her standards of accuracy and clarity.

I thank Francisco Ayala, John Beatty, Matt Chew, James Crow, George Garratty, Sandra Herbert, Richard Highton, Richard Lewontin, David O'Brochta, Anya Plutynski, David Rudge, Ezra Shahn, V. Betty Smocovitis, Carol Sokolski, Roger Thomas, Polly Winsor, and Nick Zimmermann for answering my questions and providing much valuable information in correspondence. Two anonymous referees provided extremely useful critiques.

My research was supported in part by a fellowship from the John Simon Guggenheim Memorial Foundation and by the Institute for Physical Science and Technology at the University of Maryland.

215. Creationists were not as successful in banning evolution from college biology texts. A copy of a popular textbook (Mavor 1948) obtained by interlibrary loan from Liberty Baptist College in Lynchburg, Virginia, has a sticker preceding the title page that reads:

> TO THE READER: Use of this volume as a text for reference in Liberty Baptist College is not an endorsement of its contents from the standpoint of morals, philosophy, theology, or scientific hypotheses. It is necessary to use books whose contents the college cannot wholly endorse. The position of Liberty Baptist College on the fundamentals of the faith and the separated Christian life is well known.

> Mavor, a professor of biology at Union College, accepted evolution by natural selection (de Vries' version) and gave four pieces of evidence for it, which the sticker does not attempt to refute.

....

TABLES
....

TABLE 1
Founders and leaders of the evolutionary synthesis

Theodosius **Dobzhansky** (1900–1975): MNAS, FMRS; zoology; Columbia University

Ronald Aylmer **Fisher** (1890–1962): FRS, FANAS; genetics, statistics, eugenics; University College London, Cambridge University

Edmund Briscoe **Ford** (1901–1988): FRS; genetics, zoology, comparative anatomy; Oxford University

John Burdon Sanderson **Haldane** (1892–1964): FRS; genetics and biometry; University College London, Indian Statistical Institute (Calcutta)

Julian Sorell **Huxley** (1887–1975): FRS; King's College London, Zoological Society of London

Ernst **Mayr** (1904–2005): MNAS, FMRS; ornithology; Harvard University, American Museum of Natural History (New York)

George Gaylord **Simpson** (1902–1984): MNAS, FMRS; paleontology; Harvard University, American Museum of Natural History (New York)

G. Ledyard **Stebbins Jr.** (1906–2000): MNAS; genetics, botany; University of California

Sewall (Green) **Wright** (1889–1988): MNAS, FMRS; University of Chicago, U.S. Department of Agriculture

---

Note: Limited to those with significant English-language publications, 1930–1950. FANAS = Foreign Associate of the National Academy of Sciences (United States); FMRS = Foreign Member of the Royal Society of London; FRS = Fellow of the Royal Society of London; MNAS = Member of the National Academy of Sciences (United States), elected before 1964.

TABLE 2

Factors in evolution: Textbooks, popular articles, and books

	1941–1945	1946–1950	1951–1955	1956–1960	1961–1965	1966–1970	Total
NS, small mutations (drift is insignificant in evolution or not mentioned)	11	17	14	16	27	33	118
NS, small mutations (but drift, including founder principle, is also significant in some cases)	2	7	3	9	13	26	60
Rejects NS	1	1	1	1	0	0	5
NS, including large mutations (de Vries, Goldschmidt)	15	16	10	3	2	8	56
NS + Lamarckian effects (including Baldwin effect, canalization, etc.)	4	1	1	1	1	1	10
NS + orthogenesis	6	4	1	2	1	1	15
Number of publications	29	41	28	29	42	60	229

Note: Limited to publications in English by authors who are either on the faculty of colleges or universities or who are professional staff at museums or other research institutions; not including founders and leaders, listed in table 1. NS = natural selection.

TABLE 3

*Factors in evolution: Books and technical review articles
on evolution and related topics*

	1941–1945	1946–1950	1951–1955	1956–1960	1961–1965	1966–1970	Total
NS, small mutations (drift is insignificant in evolution or not mentioned)	12	6	11	17	12	12	70
NS, small mutations (but drift, including founder principle, is also significant in some cases)	3	11	8	5	8	7	42
Rejects NS	5	1	2	1	3	2	14
NS, including large mutations (de Vries, Goldschmidt)	3	3	2	2	1	0	11
NS + Lamarckian effects (including Baldwin effect, canalization, etc.)	0	3	3	3	2	0	11
NS + orthogenesis	3	3	3	2	1	0	12
Number of publications	26	22	22	24	23	20	137

Note: Limited to publications in English by authors who are either on the faculty of colleges or universities or who are professional staff at museums or other research institutions; not including founders and leaders, listed in table 1. NS = natural selection.

TABLE 4

*Evidence for natural selection: Most frequently mentioned
in biology textbooks, popular articles, and books*

	1941–1945	1946–1950	1951–1955	1956–1960	1961–1965	1966–1970	Total
Peppered moth (Kettlewell) and other types of industrial melanism	1	1	1	7	13	29	52
Protective coloration (not including industrial melanism, peppered moth, snails)	2	6	1	5	8	9	31
Resistance to insecticides	0	2	2	3	4	9	20
Mimicry	0	0	1	2	5	6	14
Darwin's finches (Lack)	0	0	0	0	3	10	13
Resistance to antibiotics	0	2	0	0	2	9	13
Chromosome inversions and other seasonal cycles in *Drosophila, Adalia punctata,* etc.	0	3	2	2	1	5	13
Wingless insects on ocean islands	3	3	1	0	3	4	14
Number of publications	30	43	25	32	42	60	232

Note: Authors limited as in tables 2 and 3, but now including founders and leaders. Publications that rejected natural selection are not included in number of publications at bottom. Frequency = mentioned ten or more times.

TABLE 5

*Evidence for natural selection: Most frequently mentioned
in books and technical reviews on evolution*

	1941–1945	1946–1950	1951–1955	1956–1960	1961–1965	1966–1970	Total
Peppered moth (Kettlewell) and other types of industrial melanism	2	3	7	12	14	13	51
Blood groups and disease polymorphism	0	0	4	9	6	13	32
Chromosome inversions and other seasonal cycles in *Drosophila, Adalia punctata,* etc.	1	3	11	4	7	7	33
Snails (Cain and Sheppard)	0	0	3	6	8	8	25
Sickle cell and malaria	0	0	3	8	4	10	25
Mimicry	0	2	3	5	8	4	22
Resistance to insecticides	1	3	7	3	5	4	23
Protective coloration (not including industrial melanism, peppered moth, snails)	3	2	7	3	5	2	22
Resistance to antibiotics	0	2	1	1	5	4	13
Number of publications	29	24	28	38	26	27	172

Note: Limited as in tables 2 and 3, but now including founders and leaders. Publications that rejected natural selection are not included in number of publications at bottom.
Frequency = mentioned ten or more times.

BIBLIOGRAPHY

....

Adams, M. B. "The Founding of Population Genetics: Contributions of the Chetverikov School, 1924–1934." *Journal of the History of Biology* 1 (1968): 23–39
———. "Towards a Synthesis: Population Concepts in Russian Evolutionary Thought, 1925–1935." *Journal of the History of Biology* 3 (1970): 107–29.
———, ed. *The Evolution of Theodosius Dobzhansky*. Princeton, NJ: Princeton University Press, 1994.

Agar, W. E., F. H. Drummond, O. W. Tiegs, and M. M. Gunson. "Fourth (Final) Report on a Test of McDougall's Lamarckian Experiment on the Training of Rats." *Journal of Experimental Biology* 31 (1954): 307–21.

Aird, I., H. H. Bentall, and J. A. F. Roberts. "A Relationship between Cancer of Stomach and ABO Blood Groups." *British Medical Journal* 1 (1953): 799–801.

Aird, I., H. H. Bentall, J. A. Mehigan, and J. A. F. Roberts. "The Blood Groups in Relation to Peptic Ulceration and Carcinoma of the Colon, Rectum, Breast, and Bronchus: An Association between the ABO Groups and Peptic Ulceration." *British Medical Journal* 2 (1954): 315–21.

Allchin, D. "Wallowing in the Wastebin." Review of *Theories on the Scrap Heap: Scientists and Philosophers on the Falsification, Rejection, and Replacement of Theories*, by J. Losee. *Science* 311 (2006): 781–2.

Allee, W. C. *The Social Life of Animals*. New York: Norton, 1938.

Allee, W. C., A. E. Emerson, O. Park, T. Park, and K. P. Schmidt. *Principles of Animal Ecology*. Philadelphia: Saunders, 1949.

Allen, E. J. "The Origin of Adaptations." *Proceedings of the Linnaean Society of London* 141 (1930): 119–38.

Allen, G. E. "Thomas Hunt Morgan and the Problem of Natural Selection." *Journal of the History of Biology* 1 (1968):113–39.
———. *Thomas Hunt Morgan: The Man and His Science*. Princeton, NJ: Princeton University Press, 1978.
———. "Naturalists and Experimentalists: The Genotype and the Phenotype." *Studies in the History of Biology* 3 (1979): 179–209.
———. "The Evolutionary Synthesis: Morgan and Natural Selection Revisited." In *The Evolutionary Synthesis*, edited by E. Mayr and W. B. Provine, 356–84. Cambridge, MA: Harvard University Press, 1980.
———. "The Several Faces of Darwin: Materialism in Nineteenth- and Twentieth-Century Evolutionary Theory." In *Evolution from Molecules to Men*, edited by D. S. Bendall, 81–102. Cambridge: Cambridge University Press, 1983.

———. "Theodosius Dobzhansky, the Morgan Lab, and the Breakdown of the Naturalist/ Experimentalist Dichotomy, 1927–1947." In *The Evolution of Theodosius Dobzhansky*, edited by M. B. Adams, 87–98. Princeton, NJ: Princeton University Press, 1994.

Allison, A. C. "Notes on Sickle-Cell Polymorphism." *Annals of Human Genetics* 19 (1954): 39–51.

———. "Aspects of Polymorphism in Man." *Cold Spring Harbor Symposia on Quantitative Biology* 20 (1955): 239–55. Reprinted in *Evolution*, edited by G. E. Brosseau Jr., 203–26. Dubuque, IA: Brown, 1967. Abridged in *The Process of Biology: Primary Sources*, edited by J. J. W. Baker and G. E. Allen, 323–43. Reading, MA: Addison-Wesley, 1970.

———. "Genetics and Infectious Disease." In *Haldane and Modern Biology*, edited by K. R. Dronamraju, 179–201. Baltimore: Johns Hopkins University Press, 1968.

American Institute of Biological Sciences, Biological Sciences Curriculum Study. *High School Biology: BSCS Green Version*. Chicago: Rand McNally, 1963.

Anderson, E. *Plants, Man, and Life*. Berkeley: University of California Press, 1952.

Angner, E. "The History of Hayek's Theory of Cultural Evolution." *Studies in History and Philosophy of Biological and Biomedical Sciences* 33 (2002): 695–718.

Ariew, A. "Under the Influence of Malthus's Law of Population Growth: Darwin Eschews the Statistical Techniques of Adolphe Quetelet." *Studies in History and Philosophy of Biological and Biomedical Sciences* 38 (2007): 1–19.

Ariew, A., and R. C. Lewontin. "The Confusions of Fitness." *British Journal for the Philosophy of Science* 55 (2004): 347–63.

Ayala, F. J. "Evolution of Fitness in Experimental Populations of *Drosophila serrata*." *Science* 150 (1965): 903–5.

———. "Genotype, Environment, and Population Numbers." *Science* 162 (1968a): 1453–9.

———. "Biology as an Autonomous Science." *American Scientist* 56 (1968b): 207–21.

———. "Theodosius Dobzhansky: The Man and the Scientist." *Annual Review of Genetics* 10 (1976): 1–6.

———. "Philosophical Issues." Chap. 16 in *Evolution*, edited by T. Dobzhansky, F. J. Ayala, G. L. Stebbins, and J. W. Valentine. San Francisco: Freeman, 1977.

———. "Dobzhansky, Theodosius." In *Dictionary of Scientific Biography*, vol. 17, no. 2, edited by F. L. Holmes, S233–S242. New York: Charles Scribner's Sons, 1990.

Ayala, F. J., and W. M. Fitch. "Genetics and the Origin of Species." *Proceedings of the National Academy of Sciences* 94 (1997): 7691–7.

Babcock, E. B. "Genetic Evolutionary Processes." *Proceedings of the National Academy of Sciences* 20 (1934): 510–5.

Babcock, E. B., G. L. Stebbins Jr., and J. A. Jenkins. "Genetic Evolutionary Processes in *Crepis*." *American Naturalist* 76 (1942): 337–63.

Bajema, C. J., ed. *Natural Selection in Human Populations: The Measurement of Ongoing Genetic Evolution in Contemporary Societies*. New York: Wiley, 1971.

Baker, J. J. W., and G. E. Allen. *The Study of Biology*. Reading, MA: Addison-Wesley, 1967.

———, eds. *The Process of Biology: Primary Sources*. Reading, MA: Addison-Wesley, 1970.

Barash, D. P. "Nature Takes Only Tiny Steps but Still Surpasses Our Reckoning." *Chronicle of Higher Education*, 18 April 2003, sec. B.

Barker, A. D. "An Approach to the Theory of Natural Selection." *Philosophy* 44 (1969): 271–90.

Barnes, E. W. Contribution to a British Association Discussion on "The Evolution of the Universe." *Nature* 128 (1931): 719–22.

———. *Scientific Theory and Religion: The World Described by Science and Its Spiritual Interpretation.* Cambridge: Cambridge University Press, 1933.

Barnett, S. A., ed. *A Century of Darwin.* London: Heinemann, 1958.

Beatty, J. "What's Wrong with the Received View of Evolutionary Theory?" *PSA 1980* 2: 397–426. In *Proceedings of the 1980 Biennial Meeting of the Philosophy of Science Association*, vol. 2, edited by P. Asquith and R. Giere. East Lansing, MI: Philosophy of Science Association, 1981.

———. "Chance and Natural Selection." *Philosophy of Science* 51 (1984): 183–211.

———. "Dobzhansky and Drift: Facts, Values, and Chance in Evolutionary Biology." In *The Probabilistic Revolution*, vol. 2, edited by L. Krüger et al., 271–311. Cambridge, MA: MIT Press, 1987a.

———. "Natural Selection and the Null Hypothesis." In *The Latest on the Best: Essays on Evolution and Optimality*, edited by J. Dupre, 53–75. Cambridge, MA: MIT Press, 1987b.

Beaver, W. C. *General Biology.* 5th ed. St. Louis: Mosby, 1958.

———. *General Biology.* 6th ed. St. Louis: Mosby, 1962.

Beaver, W. C., and G. B. Noland. *General Biology.* 7th ed. St. Louis: Mosby, 1966.

Bennett, J. H., ed. *Natural Selection, Heredity, and Eugenics: Including Selected Correspondence of R. A. Fisher with Leonard Darwin and Others.* Oxford: Clarendon Press, 1983.

Benson, S. B. "Concealing Coloration among Some Desert Rodents of the Southwestern United States." *University of California Publications in Zoology* 40 (1933): 1–70.

Berry, R. J. *Neo-Darwinism.* London: Arnold, 1982.

Bertomeu-Sánchez, J. R., A. Garcia-Belmar, A. Lundgren, and M. Patiniotis. "Introduction: Scientific and Technological Textbooks in the European Periphery." *Science and Education* 15 (2006): 657–65.

Beurton, P. "Zur Ausbildung der synthetische Theorie der biologischen Evolution." In *Die Entstehung biologischen Disziplinen, II*, edited by U. Hossfeld and T. Junker, 231–44. Berlin: Verlag für Wissenschaft und Bildung, 2002.

Blackmun, H. A., et al. "Majority Opinion of the Supreme Court in the Case of William Daubert, *et ux., etc., et al.*, Petitioners, v. Merrell Dow Pharmaceuticals, Inc., No. 92–102. Argued March 30, 1993. Decided June 28, 1993." In *West's Supreme Court Reporter, Interim Edition, 113B: Cases Argued and Determined in the Supreme Court of the United States, October Term, 1992*, 2786–2800. St. Paul, MN: West Publishing Company, 1993.

Bonner, J. T. *The Ideas of Biology.* New York: Harper, 1962.

Bowler, P. J. "Hugo de Vries and Thomas Hunt Morgan: The Mutation Theory and the Spirit of Darwinism." *Annals of Science* 35 (1978): 55–73.

———. "Theodor Eimer and Orthogenesis: Evolution by 'Definitely Directed Variation.'" *Journal of the History of Medicine and Allied Sciences* 34 (1979): 30–73.

———. *The Eclipse of Darwinism.* Baltimore: Johns Hopkins University Press, 1983.

———. "Evolution and the Eucharist: Bishop E. W. Barnes on Science and Religion in the 1920s and 1930s." *British Journal for the History of Science* 31 (1998): 453–67.

———. *Evolution: The History of an Idea.* 3rd ed. Berkeley: University of California Press, 2003.

———. "The Spectre of Darwinism: The Popular Image of Darwinism in Early Twentieth-Century Britain." In *Darwinian Heresies*, edited by A. Lustig et al., 48–68. New York: Cambridge University Press, 2004.

Box, J. F. *R. A. Fisher: The Life of a Scientist.* New York: Wiley, 1978.

Brandon, R. "The Difference between Selection and Drift: A Reply to Millstein." *Biology and Philosophy* 20 (2005): 153–70.

Breder, C. M., Jr. "A Consideration of Evolutionary Hypotheses in Reference to the Origin of Life." *Zoologica* 27, no. 3/4 (1942): 131–43.

Brömer, R., U. Hossfeld, and N. A. Rupke, eds. *Evolutionsbiologie von Darwin bis Heute.* Berlin: Verlag für Wissenschaft und Bildung, 2000.

Brosseau, G. E., Jr., ed. *Evolution.* Dubuque, IA: Brown, 1967.

Brower, J. V. Z. "Experimental Studies of Mimicry in Some North American Butterflies." *Evolution* 12 (1958): 32–47, 123–36, 273–85.

———. "Experimental Studies of Mimicry, IV: The Reactions of Starlings to Different Proportions of Models and Mimics." *American Naturalist* 94 (1960): 271–82.

Brown, R., and J. F. Danielli, eds. *Symposia of the Society for Experimental Biology: Evolution.* No. 7 New York: Academic Press, 1953.

Brues, A. M. "Selection and Polymorphism in the A-B-O Blood Groups." *American Journal of Physical Anthropology* 12 (1954): 559–97.

Brush, S. G. "A History of Random Processes, I: Brownian Motion from Brown to Perrin." *Archive for History of Exact Sciences* 5 (1968): 1–36.

———. *The Kind of Motion We Call Heat: A History of the Kinetic Theory of Gases in the 19th Century.* Amsterdam: North-Holland Publishing Company, 1976.

———. *The Temperature of History: Phases of Science and Culture in the Nineteenth Century.* New York: Franklin, 1978.

———. "The Chimerical Cat: Philosophy of Quantum Mechanics in Historical Perspective." *Social Studies of Science* 10 (1980): 393–447.

———. "Prediction and Theory Evaluation: The Case of Light Bending." *Science* 246 (1989): 1124–9.

———. "Prediction and Theory Evaluation: Cosmic Microwaves and the Revival of the Big Bang." *Perspectives on Science* 1 (1993a): 565–602.

———. "Prediction and Theory Evaluation: Subatomic Particles." *Rivista di Storia della Scienza,* 2nd ser., 1 (1993b): 47–152.

———. "Dynamics of Theory Change: The Role of Predictions." *PSA 1994* 2: 133–45. In *Proceedings of the 1994 Biennial Meeting of the Philosophy of Science Association,* vol. 2, edited by D. Hull, M. Forbes, R. M. Burian. East Lansing, MI: Philosophy of Science Association, 1995.

———. "The Reception of Mendeleev's Periodic Law." *Isis* 87 (1996a): 595–628.

———. *Fruitful Encounters: The Origin of the Solar System and of the Moon from Chamberlin to Apollo.* New York: Cambridge University Press, 1996b.

———. *Transmuted Past: The Age of the Earth and the Evolution of the Elements from Lyell to Patterson.* New York: Cambridge University Press, 1996c.

———. "Dynamics of Theory Change in Chemistry." *Studies in History and Philosophy of Science* 30 (1999a): 21–79, 263–302.

———. "Why Was Relativity Accepted?" *Physics in Perspective* 1 (1999b): 184–214.

———. Review of *Einstein, Picasso: Space, Time, and the Beauty that Causes Havoc,* by A. I. Miller. *Physics Today* 54, no. 12 (2001): 49–50. Letters in response by C. R. Zigmund, H. Tourin, and S. G. Brush appear in *Physics Today* 55, no. 5 (2002): 12.

———. "How Theories Became Knowledge: Morgan's Chromosome Theory of Heredity in America and Britain." *Journal of the History of Biology* 35 (2002): 471–535.

———. "How Theories Became Knowledge: Why Science Textbooks Should Be Saved." In *Essays on the Research Value of Printed Materials in the Digital Age*, edited by Y. Carignan et al., 45–57. Lanham, MD: Scarecrow Press, 2005.

Brush, S. G., H. L. Sahlin, and E. Teller. "A Monte Carlo Study of a One-Component Plasma, I." *Journal of Chemical Physics* 45 (1966): 2102–18.

Buckwalter, J. A., E. B. Wohlwend, D. C. Colter, R. T. Tidrick, and L. A. Knowler. "ABO Blood Groups and Disease." *Journal of the American Medical Association* 162 (1956a): 1210–5.

———. "Peptic Ulceration and ABO Blood Groups." *Journal of the American Medical Association* 162 (1956b): 1215–20.

Buckwalter, J. A., E. B. Wohlwend, D. C. Colter, and R. T. Tidrick. "Natural Selection Associated with the ABO Blood Group." *Science* 123 (1956): 840–1.

Buican, D. "La Génétique Classique en France Devant le Néo-Lamarckisme Tardif." *Archives Internationales d'Histoire des Sciences* 33 (1983): 300–24.

———. *Histoire de la Génétique et de l'Evolutionnisme en France*. Paris: Presses Universitaires de France, 1984.

———. *La Revolution de l'Evolution: L'Evolution de l'Evolutionnisme*. Paris: Presses Universitaires de France, 1989.

Burian, R. M. "Dobzhansky on Evolutionary Dynamics: Some Questions about His Russian Background." In *The Evolution of Theodosius Dobzhansky*, edited by M. B. Adams, 129–40. Princeton, NJ: Princeton University Press, 1994.

Cain, A. J. "So-Called Non-adaptive or Neutral Characters in Evolution." *Nature* 168 (1951a): 424.

———. "Non-adaptive or Neutral Characters in Evolution." *Nature* 168 (1951b): 1049.

———. "The Perfection of Animals." In *Viewpoints in Biology*, vol. 3, edited by J. D. Carthy and C. L. Duddington, 36–63. London: Butterworths, 1964.

Cain, A. J., and P. M. Sheppard. "Selection in the Polymorphic Land Snail, *Cepaea nemoralis.*" *Heredity* 4 (1950): 275–94.

Cain, J. "A Matter of Perspective: Multiple Readings of George Gaylord Simpson's *Tempo and Mode in Evolution.*" *Archives of Natural History* 30 (2003): 28–39.

———. *Exploring the Borderlands: Documents of the Committee on Common Problems of Genetics, Paleontology, and Systematics, 1943–1944*. Vol. 94, pt. 2, *Transactions*. Philadelphia: American Philosophical Society, 2004.

Campbell, D. T. "Blind Variation and Selective Retention in Creative Thought as in Other Knowledge Processes." *Psychological Review* 67 (1960): 380–400.

———. "Variation and Selective Retention in Socio-Cultural Evolution." In *Social Change in Developing Areas: A Reinterpretation of Evolutionary Theory*, edited by H. R. Barrington, G. I. Blanksten, and R. W. Mach, 19–49. Cambridge, MA: Schenkman, 1965.

———. "Natural Selection as an Epistemological Model." In *A Handbook of Method in Cultural Anthropology*, edited by R. Naroll and R. Cohen, 51–85. New York: Natural History Press, 1970.

Carlson, E. A., ed. *Modern Biology: Its Conceptual Foundation*. New York: Braziller, 1967.

Carson, H. L. "Cytogenetics and the Neo-Darwinian Synthesis." In *The Evolutionary Synthesis*, edited by E. Mayr and W. B. Provine, 86–95. Cambridge, MA: Harvard University Press, 1980.

Carter, G. S. *Animal Evolution: A Study of Recent Views of Its Causes*. London: Sidgwick and Jackson, 1951a.

———. "Non-adaptive Characters in Evolution." *Nature* 168 (1951b): 700–1.

———. "Non-adaptive or Neutral Characters in Evolution." *Nature* 168 (1951c): 1049.

———. *Animal Evolution: A Study of Recent Views of Its Causes.* 2nd ed. London: Sidgwick and Jackson, 1954.

———. *A Hundred Years of Evolution.* London: Sidgwick and Jackson, 1957.

Cavalli-Sforza, L. L. "'Genetic Drift' in an Italian Population." *Scientific American* 221, no. 2 (1969): 30–37, 136.

Cavalli-Sforza, L. L., and A. W. F. Edwards. "Analysis of Human Evolution." *Proceedings of the Eleventh International Congress of Genetics*, 1964, 923–35.

Ceccarelli, L. *Shaping Science with Rhetoric: The Cases of Dobzhansky, Schrödinger, and Wilson.* Chicago: University of Chicago Press, 2001.

Chetverikov, S. S. "O Nekotorykh Momentakh Evolyutsionnogo Procetsessa c Tochki Zreniya Sovremennoi Genetika." *Zhurnal Eksperimentalnoi Biologii*, n.s., 2, no. 1 (1926): 3–54.

———. *On Certain Aspects of the Evolutionary Process from the Standpoint of Modern Genetics.* Edited and with an introduction by C. D. Mellon. Translated by M. Barker. Placitas, NM: Genetics Heritage Press, 1997.

Clark, P. "Atomism versus Thermodynamics." In *Method and Appraisal in the Physical Sciences*, edited by C. Howson, 41–105. New York: Cambridge University Press, 1976.

Clark, R. W. *The Huxleys.* New York: McGraw-Hill, 1968.

———. *J. B. S.: The Life and Work of J. B. S. Haldane.* New York: Coward-McCann, 1969.

Clarke, B. C. "Edmund Briscoe Ford: 23 April 1901—21 January 1988." *Biographical Memoirs of Fellows of the Royal Society of London* 41 (1995): 147–68.

Clarke, C. A. "Blood Groups and Disease." *Progress in Medical Genetics* 1 (1961): 80–119.

———. "Blood Group Interactions between Mother and Foetus." In *Ecological Genetics and Evolution*, edited by C. R. Creed, 324–44. Oxford: Blackwell, 1971.

———. "Philip MacDonald Sheppard: 27 July 1921—17 October 1976." *Biographical Memoirs of Fellows of the Royal Society of London* 23 (1977): 465–500.

Clausen, J., D. D. Keck, and W. M. Hiesey. *Experimental Studies on the Nature of Species, I: Effect of Varied Environments on Western North American Plants.* Washington, DC: Carnegie Institution of Washington, 1940.

———. "Heredity of Geographically and Ecologically Isolated Races." *American Naturalist* 81 (1947): 114–33.

Cockrum, E. L., and W. J. McCauley. *Zoology.* Philadelphia: Saunders, 1965.

Conklin, E. G. "Ends as Well as Means in Life and Evolution." *Transactions of the New York Academy of Sciences*, 2nd ser., 6 (1944): 125–136.

Cooper, J. E. "Of Microbes and Men: A Scientific Biography of René Jules Dubos." PhD diss., Rutgers University, 1998.

Cooper, R. A. "The Goal of Evolution Instruction: Should We Aim for Belief or Scientific Literacy?" *Reports of the National Center for Science Education* 21, nos. 1/2 (2001): 14–18.

Cott, H. B. *Adaptive Coloration in Animals.* London: Oxford University Press, 1940.

Coyne, J. A., N. H. Barton, and M. Torelli. "Perspective: A Critique of Sewall Wright's Shifting Balance Theory of Evolution." *Evolution* 51 (1997): 643–71.

Cravens, H. *The Triumph of Evolution: American Scientists and the Heredity-Environment Controversy.* A reprint of the 1978 edition with a new preface. Baltimore: Johns Hopkins University Press, 1988

Creager, A. N. H. "Adaptation or Selection? Old Issues and New Stakes in the Postwar

Debates over Bacterial Drug Resistance." *Studies in History and Philosophy of Biological and Biomedical Sciences* 38 (2007): 159–90.

Crombie, A. C. "Interspecific Competition." *Journal of Animal Ecology* 16 (1947): 44–73.

Crow, J. F. "Eighty Years Ago: The Beginnings of Population Genetics." *Genetics* 119 (1988): 473–6.

———. "Sewall Wright's Place in Twentieth Century Biology." *Journal of the History of Biology* 23 (1990a): 57–89.

———. "Fisher's Contribution to Genetics and Evolution." *Theoretical Population Biology* 38 (1990b): 263–75.

———. "R. A. Fisher, a Centennial View." *Genetics* 124 (1990c): 207–11. Reprinted in *Perspectives on Genetics*, edited by J. F. Crow and W. F. Dove, 142–6. Madison: University of Wisconsin Press, 2000.

———. "Was Wright Right?" *Science* 253 (1991): 973.

———. "Sixty Years Ago: The 1932 International Congress of Genetics." *Genetics* 131 (1992a): 761–8. Reprinted in *Perspectives on Genetics*, edited by J. F. Crow and W. F. and Dove, 297–301. Madison: University of Wisconsin Press, 2000.

———. "Centennial: J. B. S. Haldane, 1892–1964." *Genetics* 130 (1992b): 1–6. Reprinted in *Perspectives on Genetics*, edited by J. F. Crow and W. F. Dove, 253–8. Madison: University of Wisconsin Press, 2000.

———. "H. J. Muller's Role in Evolutionary Biology." In *The Founders of Evolutionary Genetics*, edited by S. Sarkar, 83–105. Boston: Kluwer, 1992c.

———. "Sewall Wright: December 21, 1889—March 3, 1988." *Biographical Memoirs of the National Academy of Sciences* 64 (1994): 439–69.

———. "Here's to Fisher, Additive Genetic Variance, and the Fundamental Theorem of Natural Selection." *Evolution* 56 (2002): 1313–6.

Crow, J. F., and S. Abrahamson. "Seventy Years Ago: Mutation Becomes Experimental." *Genetics* 147 (1997): 1491–6. Reprinted in *Perspectives on Genetics*, edited by J. F. Crow and W. F. Dove, 632–7. Madison: University of Wisconsin Press, 2000.

Crow, J. F., and W. F. Dove, eds. *Perspectives on Genetics: Anecdotal, Historical, and Critical Commentaries, 1987–1998*. Articles reprinted from the journal *Genetics*. Madison: University of Wisconsin Press, 2000.

Crow, J. F., W. R. Engels, and C. Denniston. "Phase Three of Wright's Shifting-Balance Theory." *Evolution* 44 (1990): 233–47.

Curtis, H. *Biology*. Special ed. New York: Worth Publishers, 1968.

Dalbiez, R. "Le Transformisme et la Philosophie." In *Le Transformisme*, edited by L. Cuenot, et al., 173–218. Paris: Vrin, 1927.

Darden, L. "Relations among Fields in the Evolutionary Synthesis." In *Integrating Scientific Disciplines*, edited by W. Bechtel, 113–23. Dordrecht, Netherlands: Nijhoff, 1986.

Darden, L., and J. A. Cain. "Selection Type Theories." *Philosophy of Science* 56 (1989): 106–29. Reprinted as chap. 8 of *Reasoning in Biological Discoveries*, edited by L. Darden. New York: Cambridge University Press, 2006.

Darlington, C. D. *Chromosome Botany and the Origins of Cultivated Plants*. 2nd ed. London: Allen and Unwin, 1963.

———. "The Evolution of Genetic Systems: Contributions of Cytology to Evolutionary Theory." In *The Evolutionary Synthesis*, edited by E. Mayr and W. B. Provine, 70–80. Cambridge, MA: Harvard University Press, 1980.

Darwin, C. R. *On the Origin of Species by Means of Natural Selection: or, The Preservation of Favoured Races in the Struggle for Life.* London: Murray, 1859.

———. *Natural Selection.* Edited by R. C. Stauffer. New York: Cambridge University Press, 1975.

———. *The Correspondence of Charles Darwin.* Vol. 5, *1851–1855.* Edited by F. Burkhardt and S. Smith. New York: Cambridge University Press, 1989.

Daston, L., and M. Otte, eds. "Style in Science: Introduction." *Science in Context* 4 (1991): 223–32.

de Beer, G. *Reflections of a Darwinian: Essays and Addresses.* London: Nelson, 1962.

Delsol, M. *L'Evolution Biologique en Vingt Propositions: Essai d'Analyse épistémologique de la Théorie Synthetique de l'Évolution.* Paris: Vrin, 1991.

Demerec, M. "Production of *Staphylococcus* Strains Resistant to Various Concentrations of Penicillin." *Proceedings of the National Academy of Sciences* 31 (1945): 16–24.

Depew, D. J., and B. H. Weber, eds. *Evolution at a Crossroads: The New Biology and the New Philosophy of Science.* Cambridge, MA: MIT Press, 1985a.

———. "Innovation and Tradition in Evolutionary Theory: An Interpretive Afterword." In *Evolution at a Crossroads,* edited by D. J. Depew and B. H. Weber, 227–60. Cambridge, MA: MIT Press, 1985b.

———. *Darwinism Evolving: Systems Dynamics and the Genealogy of Natural Selection.* Cambridge, MA: MIT Press, 1995.

Dietrich, M. R. "The Origins of the Neutral Theory of Molecular Evolution." *Journal of the History of Biology* 27 (1994): 21–59.

———. "Richard Goldschmidt's 'Heresies' and the Evolutionary Synthesis." *Journal of the History of Biology* 28 (1995): 431–61.

Divall, C. "From a Victorian to a Modern: Julian Huxley and the English Intellectual Climate." In *Julian Huxley: Biologist and Statesman of Science,* edited by C. K. Waters and A. Van Helden, 31–48. Houston, TX: Rice University Press, 1992.

Diver, C. "The Problem of Closely Related Species Living in the Same Area." In *The New Systematics,* edited by J. Huxley, 303–28. Oxford: Clarendon Press, 1940.

Dobzhansky, T. "Geographical Variation in Lady-Beetles." *American Naturalist* 67 (1933): 97–126.

———. "A Critique of the Species Concept in Biology." *Philosophy of Science* 2 (1935): 344–55.

———. "Studies on Hybrid Sterility, II: Localization of Sterility Factors in *Drosophila pseudoobscura* Hybrids." *Genetics* 21 (1936): 113–35.

———. *Genetics and the Origin of Species.* New York: Columbia University Press, 1937.

———. "Discovery of a Predicted Gene Arrangement in *Drosophila azteca.*" *Proceedings of the National Academy of Sciences* 27 (1941a): 47–50.

———. *Genetics and the Origin of Species.* 2nd ed. New York: Columbia University Press, 1941b.

———. "Biological Adaptation." *Scientific Monthly* 55 (1942): 391–402.

———. "Genetics of Natural Populations, IX: Temporal Changes in the Compositions of Populations of *Drosophila pseudoobscura.*" *Genetics* 28 (1943): 162–86.

———. "Genetics of Natural Populations, XIII: Recombination and Variability in Populations of *Drosophila pseudoobscura.*" *Genetics* 31 (1946): 269–90.

———. "Adaptive Changes Induced by Natural Selection in Wild Populations of *Drosophila.*" *Evolution* 1 (1947): 1–16.

———. "Mendelian Populations and Their Evolution." *American Naturalist* 84 (1950a): 401–18.

———. "The Genetic Basis of Evolution." *Scientific American* 182, no. 1 (1950b): 32–41.

———. *Genetics and the Origin of Species.* 3rd ed. New York: Columbia University Press, 1951.

———. *Evolution, Genetics, and Man.* New York: Wiley, 1955.

———. "Evolution at Work." *Science* 127 (1958): 1091–8.

———. "Variation and Evolution." *Proceedings of the American Philosophical Society* 103 (1959a): 252–63.

———. Discussion remark. *Cold Spring Harbor Symposia on Quantitative Biology* 24 (1959b): 85–86.

———. "Man and Natural Selection." *American Scientist* 49 (1961): 285–99. Reprinted in *Natural Selection in Human Populations*, edited by C. J. Bajema, 4–18. New York: Wiley, 1971.

———. *Mankind Evolving.* New Haven: Yale University Press, 1962.

———. "On Some Fundamental Concepts of Darwinian Biology." *Evolutionary Biology* 2 (1968): 1–34. Another version published with the title "On Cartesian and Darwinian Aspects of Biology: Are They Compatible?" *Graduate Journal* 8 (1968): 99–117.

———. *Genetics of the Evolutionary Process.* New York: Columbia University Press, 1970.

———. "Evolutionary Oscillations in *Drosophila pseudoobscura*." In *Ecological Genetics and Evolution*, edited by R. Creed, 109–33. Oxford: Blackwell, 1971.

———. *Dobzhansky's Genetics of Natural Populations, I–XLIII.* Edited by R. C. Lewontin, J. A. Moore, W. B. Provine, and B. Wallace. New York: Columbia University Press, 1981.

Dobzhansky, T., and M. F. A. Montagu. "Natural Selection and the Mental Capacities of Mankind." *Science* 105 (1947): 587–90.

Dobzhansky, T., and O. Pavlovsky. "An Experimental Study of Interaction between Genetic Drift and Natural Selection." *Evolution* 11 (1957): 311–9. Reprinted in *Evolution*, edited by G. E. Brosseau Jr., 249–59. Dubuque, IA: Brown, 1967.

Dobzhansky, T., and N. P. Spassky. "Genetic Drift and Natural Selection in Experimental Populations of *Drosophila pseudoobscura*." *Proceedings of the National Academy of Sciences* 48 (1962): 148–56.

Dobzhansky, T., and A. H. Sturtevant. "Inversions in the Chromosomes of *Drosophila pseudoobscura*." *Genetics* 23 (1938): 28–64.

Dodson, E. O. *A Textbook of Evolution.* Philadelphia: Saunders, 1952.

Dowdeswell, W. H. *The Mechanism of Evolution.* 2nd ed. London: Heinemann, 1958.

———. *Animal Ecology.* 2nd ed. London: Methuen, 1959.

———. *The Mechanism of Evolution.* 3rd ed. London: Heinemann, 1963.

———. "Ecological Genetics and Biology Teaching." In *Ecological Genetics and Evolution*, edited by R. Creed, 363–78. Oxford: Blackwell, 1971.

Dowdeswell, W. H., and E. B. Ford. "The Influence of Isolation on Variability in the Butterfly *Maniola jurtina* L." In *Evolution: Symposia of the Society for Experimental Biology*, no. 7, edited by R. Brown and J. F. Danielli, 254–73. New York: Academic Press, 1953.

Drabble, M. *The Peppered Moth.* New York: Harcourt, 2001.

Dronamraju, K. R. *If I Am To Be Remembered: The Life and Work of Julian Huxley, with Selected Correspondence.* River Edge, NJ: World Scientific, 1993.

Dubinin, N. P. "Genetiko-avtomaticheskie Protsessy i ikh Znachenie dlia Mekhanizma Organicheskoi Evoliutsii." *Zhurnal Eksperimental'noe Biologii* 7 (1931): 463–79.

Dubinin, N. P., and D. D. Romashov. "Geneticheskoe Stroenie Vida I ego Evoliutsiia, 1: Genetiko-avtomaticheskie Protsessy I Problema Ekogenotipov." *Biologicheskii Zhurnal* 1, nos. 5/6 (1932): 52–95.

Dubinin, N. P., and G. G. Tiniakov. "Seasonal Cycles and the Concentration of Inversions in Populations of *Drosophila funebris.*" *American Naturalist* 79 (1945): 570–2.

———. "Inversion Gradients and Natural Selection in Ecological Races of *Drosophila funebris.*" *Genetics* 31 (1946a): 537–45.

———. "Structural Chromosome Variability in Urban and Rural Populations of *Drosophila funebris.*" *American Naturalist* 80 (1946b): 393–6.

———. "Seasonal Cycle and Inversion Frequency in Populations." *Nature* 157 (1946c): 23–24.

———. "Natural Selection in Experiments with Population Inversions." *Journal of Genetics* 48 (1947a): 11–15.

———. "Inversion Gradients and Selection in Ecological Races of *Drosophila funebris.*" *American Naturalist* 81 (1947b): 148–53.

Dubos, R. J. "Microbiology." *Annual Review of Biochemistry* 11 (1942): 659–78.

———. "Trends in the Study and Control of Infectious Diseases." *Proceedings of the American Philosophical Society* 88 (1944): 208–13.

———. *The Bacterial Cell in Relation to Problems of Virulence, Immunity and Chemotherapy.* Cambridge, MA: Harvard University Press, 1945.

Dunn, E. R. "The Survival Value of Specific Characters." *Copeia*, no. 2 (1935): 85–98.

Dunn, L. C. *Heredity and Evolution in Human Populations.* Cambridge, MA: Harvard University Press, 1959.

———. *Heredity and Evolution in Human Populations.* Rev. ed. New York: Atheneum, 1967.

Dunn, L. C., and W. Landauer. "The Genetics of the Rumpless Fowl with Evidence of a Case of Changing Dominance." *Journal of Genetics* 29 (1934): 217–43

East, E. M. "Genetic Reactions in Nicotiana, III: Dominance." *Genetics* 20 (1935): 443–51.

Edwards, A. W. F. "The Genetical Theory of Natural Selection." *Genetics* 154 (2000): 1419–26.

Eldredge, N. *Unfinished Synthesis: Biological Hierarchies and Modern Evolutionary Thought.* Oxford: Oxford University Press, 1985.

Elliott, A. M. *Zoology.* New York: Appleton-Century-Crofts, 1952.

———. *Zoology.* 2nd ed. New York: Appleton-Century-Crofts, 1957.

———. *Zoology.* 3rd ed. New York: Appleton-Century-Crofts, 1963.

———. *Zoology.* 4th ed. New York: Appleton-Century-Crofts, 1968.

Elliott, A. M., and C. Ray Jr. *Biology.* 2nd ed. New York: Appleton, 1965.

Emerson, A. E. "Taxonomic Categories and Population Genetics." *Entomological News* 56, no. 1 (1945): 14–19.

Endler, J. A. *Natural Selection in the Wild.* Princeton, NJ: Princeton University Press, 1986.

———. "Natural Selection: Current Usages." In *Keywords in Evolutionary Biology*, edited by E. F. Keller and E. A. Lloyd, 220–4. Cambridge, MA: Harvard University Press, 1992.

Epling, C. *Contributions to the Genetics, Taxonomy, and Ecology of Drosophila pseudoobscura and Its Relatives, III: The Historical Background.* Washington, DC: Carnegie Institution of Washington, 1944.

Epling, C., and T. Dobzhansky. "Genetics of Natural Populations, VI: Microgeographic Races in *Linanthus parryae.*" *Genetics* 27 (1942): 317–32.

Ewens, W. J. "Beanbag Genetics and After." In *Human Population Genetics: A Centennial Tribute to J. B. S. Haldane,* edited by P. P. Majumdar, 7–29. New York: Plenum, 1993.

Fabergé, A. C. "Genetics of the *Scapiflora* Section of *Papaver,* II: The Alpine Poppy." *Journal of Genetics* 45 (1943): 139–70.

Falconer, D. S. *Introduction to Quantitative Genetics,* 2nd ed. London: Longman, 1981.

Fisher, R. A. "The Correlation between Relatives on the Supposition of Mendelian Inheritance." *Transactions of the Royal Society of Edinburgh* 52 (1918): 399–433.

———. Review of *The Relative Value of the Processes Causing Evolution,* by A. L. Hagedoorn and A. C. Hagedoorn. *Eugenics Review* 13 (1921): 467–70.

———. "On the Dominance Ratio." *Proceedings of the Royal Society of Edinburgh* 42 (1922): 321–41.

———. "On Some Objections to Mimicry Theory: Statistical and Genetic." *Transactions of the Royal Entomological Society of London* 75 (1927): 269–78.

———. *The Genetical Theory of Natural Selection.* Oxford: Oxford University Press, 1930a.

———. "Mortality amongst Plants and Its Bearing on Natural Selection." *Nature* 125 (1930b): 972–3.

———. "Indeterminism and Natural Selection." *Philosophy of Science* 1 (1934): 99–117.

———. "Has Mendel's Work Been Rediscovered?" *Annals of Science* 1 (1936a): 115–37.

———. "The Measurement of Selective Intensity." Contribution to the "Discussion on the Present State of the Theory of Natural Selection." *Proceedings of the Royal Society of London* B121 (1936b): 58–62.

———. "Dominance in Poultry: Feathered Feet, Rose Comb, Internal Pigment and Pile." *Proceedings of the Royal Society of London* B125 (1938): 25–48.

———. "Average Excess and Average Effect of a Gene Substitution." *Annals of Eugenics* 11 (1941): 53–63.

———. "The Renaissance of Darwinism." *Listener* 37 (1947): 1001.

———. *Creative Aspects of Natural Law.* Cambridge: Cambridge University Press, 1950.

———. "Population Genetics." Contribution to the Croonian Lecture, 11 June 1953. *Proceedings of the Royal Society of London* B141 (1953): 510–23.

———. "Retrospect of the Criticisms of the Theory of Natural Selection." In *Evolution as a Process,* edited by J. S. Huxley, 84–98. London: Allen and Unwin, 1954.

———. *The Genetical Theory of Natural Selection.* 2nd ed. New York: Dover, 1958.

———. *Collected Papers of R. A. Fisher.* Edited by J. H. Bennett. 5 vols. Adelaide, Australia: University of Adelaide, 1971–1974.

———. *The Genetical Theory of Natural Selection: A Complete Variorum Edition.* Edited and with foreword and notes by J. H. Bennett. Oxford: Oxford University Press, 1999.

Fisher, R. A., and E. B. Ford. "Variability of Species." *Nature* 148 (1926): 515–6.

———. "The Variability of Species in the *Lepidoptera,* with Reference to Abundance and Sex." *Transactions of the Royal Entomological Society of London* 76 (1928): 367–79.

———. "The Spread of a Gene in Natural Conditions in a Colony of the Moth *Panaxia dominula* L." *Heredity* 1 (1947): 143–74.

———. "The 'Sewall Wright Effect.'" *Heredity* 4 (1950): 117–9.

Fisher, R. A., and C. S. Stock. "Cuénot on Preadaptation." *Eugenics Review* 7 (1915): 46–61.

Flew, A. G. N. "The Concept of Evolution: A Comment." *Philosophy* 41 (1966): 70–75.

Ford, E. B. *Mendelism and Evolution.* London: Methuen, 1931.

———. "The Genetics of Mimicry." In *Mimicry*, edited by G. D. H. Carpenter, 103–24. London: Methuen, 1933.

———. "Problems of Heredity in the *Lepidoptera*." *Biological Reviews* 12 (1937): 461–503.

———. "Polymorphism and Taxonomy." In *The New Systematics*, edited by J. Huxley, 493–513. Oxford: Clarendon Press, 1940a.

———. "Genetic Research in the *Lepidoptera*." *Annals of Eugenics* 10 (1940b): 227–52.

———. "Polymorphism." *Biological Review* 20 (1945): 73–88.

———. "Rapid Evolution and the Conditions Which Make it Possible." *Cold Spring Harbor Symposia on Quantitative Biology* 20 (1955): 230–8.

———. *Mendelism and Evolution.* 6th ed. London: Methuen, 1957a.

———. "Polymorphism in Plants, Animals and Man." *Nature* 180 (1957b): 1315–9.

———. "Evolution in Progress." In *Evolution after Darwin*, vol. 1, edited by S. Tax, 181–96. Chicago: University of Chicago Press, 1960.

———. *Ecological Genetics.* London: Methuen, 1964.

———. *Ecological Genetics.* 3rd ed. London: Chapman and Hall, 1971.

———. *Ecological Genetics.* 4th ed. London: Chapman and Hall, 1975.

———. "Some Recollections Pertaining to the Evolutionary Synthesis." In *The Evolutionary Synthesis*, edited by E. Mayr and W. B. Provine, 334–42. Cambridge, MA: Harvard University Press, 1980.

———. "R. A. Fisher: An Appreciation." Transcript of an audiotape recorded shortly after Fisher's death in 1962. *Genetics* 171 (2005): 415–7.

Forman, P. "Weimar Culture, Causality, and Quantum Theory, 1918–1927: Adaptation by German Physicists and Mathematicians to a Hostile Intellectual Environment." *Historical Studies in the Physical Sciences* 3 (1971): 1–115.

Franklin, A., A. W. F. Edwards, D. J. Fairbanks, D. L. Hartl, and T. Seidenfeld. 2008. *Ending the Mendel-Fisher Controversy.* Pittsburgh: University of Pittsburgh Press.

Futuyma, D. J. *Evolutionary Biology.* Sunderland, MA: Sinauer, 1979.

Garber, E., S. G. Brush, and C. W. F. Everitt, eds. *Maxwell on Molecules and Gases.* Cambridge, MA: MIT Press, 1986.

Gardner, E. J. *Mechanics of Organic Evolution.* Logan: Utah State University Press, 1962.

Garratty, G. "Do Blood Groups Have a Biological Role?" In *Immunobiology of Transfusion Medicine*, 201–55. New York: Dekker, 1994.

———. "Association of Blood Groups and Disease: Do Blood Group Antigens and Antibodies Have a Biological Role?" *History and Philosophy of Life Sciences* 18 (1996): 321–44.

———. "Blood Groups and Disease: A Historical Perspective." *Transfusion Medicine Reviews* 4 (2000): 291–301.

Gayon, J. "Neo-Darwinism." In *Concepts, Theories and Rationality in the Biological Sciences*, edited by G. Wolters and J. G. Lennox, 1–25. Konstanz, Germany: Universitätsverlag; Pittsburgh: University of Pittsburgh Press, 1993.

———. *Darwinism's Struggle for Survival: Heredity and the Hypothesis of Natural Selection.* Translated by M. Gibb. New York: Cambridge University Press, 1998.

Gayon, J., and M. Veuille. "The Genetics of Experimental Populations: L'Héritier and Teissier's Population Cages." In *Thinking about Evolution*, vol. 2, edited by R. S. Singh et al., 77–102. New York: Cambridge University Press, 2001.

. . . .

Gershenson, S. "Evolutionary Studies on the Distribution and Dynamics of Melanism in the Hamster (*Cricetus cricetus* L.)." *Genetics* 30 (1945): 207–32, 233–51.

Gerson, E. M. "The American Style of Research: Evolutionary Biology, 1890–1950." PhD diss., University of Chicago, 1998.

Ghiselin, M. *The Triumph of the Darwinian Method*. Berkeley: University of California Press, 1969.

Gigerenzer, G., Z. Swijtink, T. Porter, L. Daston, J. Beatty, and L. Krüger. *The Empire of Chance: How Probability Changed Science and Everyday Life*. New York: Cambridge University Press, 1989.

Glass, B. "Genetic Changes in Human Populations, Especially Those Due to Gene Flow and Genetic Drift." *Advances in Genetics* 6 (1954): 95–139.

———. "The Biochemistry of Human Heredity." In *Frontiers of Modern Biology*, edited by G. B. Moment, 83–93. Boston: Houghton Mifflin, 1962.

———. "Timofeeff-Ressovsky, Nikolai Vladimirovich." In *Dictionary of Scientific Biography*, vol. 18, edited by F. L. Holmes, S919–S926. New York: Charles Scribner's Sons, 1990.

Glass, B., M. S. Sacks, E. Jahn, and C. Hess. "Genetic Drift in a Religious Isolate: An Analysis of the Causes of Variation in Blood Group and Other Gene Frequencies in a Small Population." *American Naturalist* 86 (1952): 145–59.

Glick, T. F., ed. *The Comparative Reception of Darwinism*. Chicago: University of Chicago Press, 1974. Reprinted with a new preface. Chicago: University of Chicago Press, 1988.

Glick, T. F., and M. G. Henderson. "The Scientific and Popular Receptions of Darwin, Freud, and Einstein." In *The Reception of Darwinism in the Iberian World: Spain, Spanish America, and Brazil*, edited by T. F. Glick, M. A. Puig-Samper, and R. Ruiz, 229–38. Dordrecht, Netherlands: Kluwer, 2001.

Glymour, B. "Wayward Modeling: Population Genetics and Natural Selection." *Philosophy of Science* 73 (2006): 369–89.

Goldschmidt, R. "Lymantria." *Bibliographia Genetica* 11 (1934): 1–186.

———. *The Material Basis of Evolution*. New Haven, CT: Yale University Press, 1940.

Goudge, T. A. *The Ascent of Life: A Philosophical Study of the Theory of Evolution*. Toronto: University of Toronto Press, 1961.

Gould, S. J. "Allometry and Size in Ontogeny and Phylogeny." *Biological Reviews* 41 (1966): 587–640.

———. "G. G. Simpson, Paleontology, and the Modern Synthesis." In *The Evolutionary Synthesis*, edited by E. Mayr and W. B. Provine, 153–72. Cambridge, MA: Harvard University Press, 1980.

———. Introduction. In reprint of *Genetics and the Origin of Species*, by T. Dobzhanzky. New York: Columbia University Press, 1982.

———. "The Hardening of the Modern Synthesis." In *Dimensions of Darwinism*, edited by M. Grene, 71–93. Cambridge: Cambridge University Press, 1983.

———. Foreword. In *Basic Questions in Paleontology: Geologic Time, Organic Evolution, and Biological Systematics*, by O. Schindewolf. Chicago: University of Chicago Press, 1993.

———. *The Structure of Evolutionary Theory*. Cambridge, MA: Harvard University Press, 2002.

Grabiner, J. V., and P. D. Miller. "Effects of the Scopes Trial." *Science* 185 (1974): 832–7.

Greene, J. C. "The History of Ideas Revisited." *Revue de Synthèse* 4 (1986): 201–27.

———. "The Interaction of Science and World View in Sir Julian Huxley's Evolutionary Biology." *Journal of the History of Biology* 23 (1990): 39–55.

———. "Science, Philosophy, and Metaphor in Ernst Mayr's Writings." *Journal of the History of Biology* 27 (1994): 311–47.

———. *Debating Darwin: Adventures of a Scholar*. Claremont, CA: Regina Books, 1999.

Greene, J. C., and M. Ruse, eds. "Ernst Mayr at Ninety." *Biology and Philosophy* 9 (1994): 253–427.

———. "On the Nature of the Evolutionary Process: The Correspondence between Theodosius Dobzhansky and John C. Greene." *Biology and Philosophy* 11 (1996): 445–91.

Gregory, W. K. "Award of the Daniel Giraud Elliot Medal (of the National Academy of Sciences, for the Year 1941)." Includes a note from A. H. Sturtevant. *American Naturalist* 80 (1946): 27–29.

Grene, M., ed. *Dimensions of Darwinism: Themes and Counterthemes in Twentieth-Century Evolutionary Theory*. Cambridge: Cambridge University Press; Paris: Editions de la Maison des Sciences de l'Homme, 1983.

Grene. M., and D. Depew. *The Philosophy of Biology: An Episodic History*. New York: Cambridge University Press, 2004.

Grimoult, C. *Histoire de l'Évolutionnisme Contemporain en France, 1945–1995*. Geneva: Librarie Droz, 2000.

———. *L'Évolution Biologique en France: Une Révolution scientifique, politique et culturelle*. Geneva: Droz, 2001.

Guterman, L. "Harvard's Ernst Mayr, a Pioneer in Evolutionary Biology, Dies at 100." *Chronicle of Higher Education*, 18 February 2005, sec. A, p. 17.

Haffer, J. "Ernst Mayr: Ornithologist, Evolutionary Biologist, Historian, and Philosopher of Science." In *Die Entstehung biologischer Disziplinen, II*, edited by U. Hossfeld and T. Junker, 125–32. Berlin: Verlag für Wissenschaft und Bildung, 2002.

Hagedoorn, A. L., and A. C. Hagedoorn. *The Relative Value of the Processes Causing Evolution*. The Hague, Netherlands: Martinus Nijhoff, 1921.

Hagen, J. B. "Experimental Taxonomy, 1930–1950: The Impact of Cytology, Ecology and Genetics on Ideas of Biological Classification." PhD thesis, Oregon State University, 1982.

———. "Experimentalists and Naturalists in Twentieth-Century Botany: Experimental Taxonomy, 1920–1950." *Journal of the History of Biology* 17 (1984): 249–70.

———. "Retelling Experiments: H. B. D. Kettlewell's Studies of Industrial Melanism in Peppered Moths." *Biology and Philosophy* 14 (1999): 39–54.

Haldane, J. B. S. "A Mathematical Theory of Natural and Artificial Selection, Part I." *Transactions of the Cambridge Philosophical Society* 23 (1924): 19–41.

———. Review of *The Genetical Theory of Natural Selection*, by R. A. Fisher *Mathematical Gazette* 15 (1931): 474–5.

———. *The Inequality of Man, and Other Essays*. London: Chatto and Windus, 1932a.

———. *The Causes of Evolution*. London: Longmans, 1932b.

———. *Marxist Philosophy and the Sciences*. New York: Random House, 1939.

———. *Adventures of a Biologist*. 3rd ed. Published in England under the title *Keeping Cool*. New York: Harper, 1940.

———. Foreword. In *Evolution*, edited by R. Brown and J. F. Danielli. New York: Academic Press, 1953.

———. "The Status of Evolution." In *Evolution as a Process*, edited by J. S. Huxley, A. C. Hardy, and E. B. Ford, 109–21. London: Allen and Unwin, 1954.

————. "The Theory of Evolution, before and after Bateson." *Journal of Genetics* 56 (1958): 11–27.

————. "Natural Selection." In *Darwin's Biological Work*, edited by P. R. Bell, 101–50. Cambridge: Cambridge University Press, 1959.

Haldane, J. B. S., and J. S. Huxley. *Animal Biology*. Oxford: Clarendon Press, 1928.

Hall, B. "Organic Selection: Proximate Environmental Effects on the Evolution of Morphology and Behaviour." *Biology and Philosophy* 16 (2001): 215–37.

Hall, N. S. "R. A. Fisher and His Advocacy of Randomization." *Journal of the History of Biology* 40 (2007): 295–325.

Hapgood, F. "The Importance of Being Ernst." *Science* 84, no. 5 (1984): 40–46.

Hardin, G. *Biology: Its Human Implications*. 2nd ed. San Francisco: Freeman, 1952.

————. *Nature and Man's Fate*. New York: Holt, Rinehart and Winston, 1959.

————. *Biology: Its Principles and Implications*. 2nd ed. San Francisco: Freeman, 1966.

————. "The Tragedy of the Commons." *Science* 162 (1968): 1243–8.

Harlan, H. V., and M. L. Martini. "The Effect of Natural Selection in a Mixture of Barley Varieties." *Journal of Agricultural Research* 57 (1938): 189–99.

Harman, O. S. *The Man Who Invented the Chromosome: The Life of Cyril Darlington*. Cambridge, MA: Harvard University Press, 2004.

Harrington, A. *Reenchanted Science: Holism in German Culture from Wilhelm II to Hitler*. Princeton, NJ: Princeton University Press, 1996.

Hartmann, M. *Die Methodologische Grundlagen der Biologie*. Leipzig, Germany: Meiner, 1933.

Harwood, J. "Geneticists and the Evolutionary Synthesis in Interwar Germany." *Annals of Science* 42 (1985): 279–301.

————. *Styles of Scientific Thought: The German Genetics Community, 1900–1933*. Chicago: University of Chicago Press, 1993.

————. "Metaphysical Foundations of the Evolutionary Synthesis: A Historiographical Note." *Journal of the History of Biology* 27 (1994): 1–20.

————. Review of *Evolutionsbiologie von Darwin bis Heute*, edited by Brömer et al. *British Journal for the History of Science* 35 (2002): 368–70.

Hecht, M. K., and W. C. Steere, eds. *Essays in Evolution and Genetics in Honor of Theodosius Dobzhansky*. New York: Appleton-Cetury-Crofts, 1970.

Heitz, E., and H. Bauer. "Beweise für die Chromosomennatur der Kernschleifen in den Knäuelkernen von *Bibio hortulans*." *Zeitschrift für Zellforschung und Mikroskopische Anatomie* 17 (1933): 67–83.

Heyes, C., and D. L. Hull, eds. *Selection Theory and Social Construction*. Albany: State University of New York Press, 2001.

Hill, W. G. "Sewall Wright: 21 December 1889—3 March 1988." *Biographical Memoirs of Fellows of the Royal Society* 36 (1990): 569–76.

Hitchcock, C., and E. Sober. "Prediction versus Accommodation and the Risk of Overfitting." *British Journal for the Philosophy of Science* 55 (2004): 1–34.

Hodge, M. J. S. "Natural Selection as a Causal, Empirical, and Probabilistic Theory." In *The Probabilistic Revolution*. Vol. 2, *Ideas in the Sciences*, edited by L. Krüger, G. Gigerenzer, and M. S. Morgan, 233–70. Cambridge, MA: MIT Press, 1987.

————. "Biology and Philosophy (Including Ideology): A Study of Fisher and Wright." In *The Founders of Evolutionary Genetics*, edited by S. Sarkar, 231–93. Dordrecht, Netherlands: Kluwer, 1992.

Holman, R. M., and W. W. Robbins. *Textbook of General Biology.* 3rd ed. New York: Wiley, 1934.

Holmes, S. J. "The Principle of Stability as a Cause of Evolution: A Review of Some Theories." *Quarterly Review of Biology* 23 (1948a): 324–32.

———. "What Is Natural Selection?" *Scientific Monthly* 67 (1948b): 324–30.

Holton, G. *Thematic Origins of Scientific Thought: Kepler to Einstein.* Rev.ed. Cambridge, MA: Harvard University Press, 1988.

Holton, G., and S. G. Brush. *Physics, the Human Adventure: From Copernicus to Einstein and Beyond.* New Brunswick, NJ: Rutgers University Press, 2001.

Hooper, J. *Of Moths and Men: The Untold Story of Science and the Peppered Moth.* New York: Norton, 2002.

Hossfeld, U. "Staatsbiologie, Rassenkunde und Moderne Synthese in Deutschland während der NS-Zeit." In *Evolutionsbiologie von Darwin bis Heute,* edited by R. Brömer, U. Hossfeld, and N. A. Rupke, 249–305. Berlin: Verlag für Wissenschaft und Bildung, 2000.

Hossfeld, U., and T. Junker, eds. *Die Entstehung biologischer Disziplinen, II: Beiträge zur 10. Jahrestagung der DGGTB in Berlin, 2001.* Berlin: Verlag für Wissenschaft und Bildung, 2002.

Hovanitz, W. "Polymorphism and Evolution." In *Evolution,* edited by R. Brown and J. F. Danielli, 238–53. New York: Academic Press, 1953.

Howe, E. M. "Untangling Sickle-Cell Anemia and the Teaching of Heterozygote Protection." *Science and Education* 16 (2007): 1–19.

Hull, D. L. *Philosophy of Biological Science.* Englewood Cliffs, NJ: Prentice-Hall, 1974.

———. *Science and Selection: Essays on Biological Evolution and the Philosophy of Science.* New York: Cambridge University Press, 2001.

Hull, D. L., and M. Ruse, eds. *The Philosophy of Biology.* Oxford: Oxford University Press, 1998.

Hull, D. L., P. D. Tessner, and A. M. Diamond. "Planck's Principle." *Science* 202 (1978): 717–23.

Huxley, J. *The Stream of Life.* London: Watts, 1926.

———. "Natural Selection and Evolutionary Progress." Presidential address to the zoology section. In *Report of the 106th Meeting of the British Association for the Advancement of Science,* 81–100. London: Office of the British Association, 1936.

———.ed. *The New Systematics.* Oxford: Clarendon Press, 1940.

———. *Evolution: The Modern Synthesis.* New York: Harper, 1942.

———. "Evolution in Action." Review of *Systematics and the Origin of Species,* by E. Mayr. *Nature* 151 (1943): 347–8.

———. "Genetics and Major Evolutionary Change." Review of *Tempo and Mode in Evolution,* by G. G. Simpson. *Nature* 156 (1945a): 3–4.

———. "Evolutionary Biology and Related Subjects." *Nature* 156 (1945b): 254–6.

———. "Species and Evolution." *Endeavour* 5, no. 17 (1946): 3–12.

———. "The Vindication of Darwinism." In *Evolution and Ethics, 1893–1943,* edited by T. H. Huxley and J. Huxley, 153–76. Reprinted from the journal *Rationalist Annual,* 1946. London: Pilot Press, 1947.

———. "Genetics, Evolution and Human Destiny." In *Genetics in the 20th Century,* edited by L. C. Dunn, 591–621. New York: Macmillan, 1951.

———. "Evolution's Copycats." *Life* 32, no. 6 (1952): 67–76. Reprinted in *New Bottles for New Wine,* by J. Huxley, 137–54. New York: Harper, 1957.

———. "The Evolutionary Process." In *Evolution as a Process*, edited by J. Huxley, A. C. Hardy, and E. B. Ford, 1–23. London: Allen and Unwin, 1954.

———. "Morphism and Evolution." *Heredity* 9 (1955): 1–52.

———. *New Bottles for New Wine*. New York: Harper, 1957.

———. Introduction. In *Evolution:The Modern Synthesis*. 2nd ed. London: Allen and Unwin, 1963.

Huxley, T. H. "Evolution and Ethics." In *Evolution and Ethics, 1893–1943*, edited by T. H. Huxley and J. Huxley, 60–102. From the Romanes Lecture, 1893. London: Pilot Press, 1947.

Jepsen, G. L. "Selection, 'Orthogenesis,' and the Fossil Record." *Proceedings of the American Philosophical Society* 93 (1949): 479–500.

Jepsen, G. L., E. Mayr, and G. G. Simpson, eds. *Genetics, Paleontology, and Evolution.* Princeton, NJ: Princeton University Press, 1949.

Johnson, K. "Ernst Mayr, Karl Jordan, and the History of Systematics." *History of Science* 43 (2005): 1–35.

Johnson, P. E. *Darwin on Trial.* 2nd ed. Downers Grove, IL: Intervarsity Press, 1993.

Junker, T. "Eugenik, Synthetische Theorie und Ethik: Der Fall Timoféeff-Ressovsky im Internationalen Kontext." In *Ethik der Biowissenschaften*, edited by E.-M. Engels et al., 7–40. Berlin: Verlag für Wissenschaft und Bildung, 1998.

———. "Synthetische Theorie, Eugenik und NS-Biologie." In *Evolutionsbiologie von Darwin bis heute*, edited by R. Brömer, U. Hossfeld, and N. A. Rupke, 307–60. Berlin: Verlag für Wissenschaft und Bildung, 2000.

———. "Darwinismus oder Synthetische Evolutionstheorie?" In *Die Entstehung biologischer Disziplinen, II*, edited by U. Hossfeld and T. Junker, 209–30. Berlin: Verlag für Wissenschaft und Bildung, 2002.

———. *Die Zweite Darwinsche Revolution: Geschichte des Synthetische Darwinismus in Deutschland, 1924 bis 1950.* Marburg: Basilisken-Presse, 2004.

Junker, T., and E.-M. Engels, eds. *Die Entstehung der synthetische Theorie: Beiträge zur Geschichte der Evolutionsbiologie in Deutschland, 1930–1950.* Berlin: Verlag der Wissenschaft und Bildung, 1999.

Junker, T., and U. Hossfeld. "Synthetische Theorie und 'Deutsche Biologie': Einführenden Essay." In *Evolutionsbiologie von Darwin bis Heute*, edited by R. Brömer, U. Hossfeld, and N. A. Rupke, 231–48.Berlin: Verlag für Wissenschaft und Bildung, 2000.

———. "The Architects of the Evolutionary Synthesis in National Socialist Germany: Science and Politics." *Biology and Philosophy* 17 (2002): 223–49.

Kemp, W. B. "Natural Selection within Plant Species as Exemplified in a Permanent Pasture." *Journal of Heredity* 28 (1937): 329–33.

Kerr, J. G. *Evolution.* London: Macmillan, 1926.

Kerr, W. E., and S. Wright. "Experimental Studies of the Distribution of Gene Frequencies in Very Small Populations of *Drosophila melanogaster*, I: Forked." *Evolution* 8 (1954a): 172–7.

———. "Experimental Studies of the Distribution of Gene Frequencies in Very Small Populations of *Drosophila melanogaster*, III: Aristapedia and Spineless." *Evolution* 8 (1954b): 293–302.

Kettlewell, H. B. D. "Selection Experiments on Industrial Melanism in the *Lepidoptera*." *Heredity* 9 (1955): 323–42.

———. "Further Selection Experiments on Industrial Melanism in the *Lepidoptera*." *He-*

redity 10 (1956): 287–301. Reprinted in *Evolution*, edited by G. E. Brosseau Jr., 228–47. Dubuque, IA: Brown, 1967.

———. "Industrial Melanism in the *Lepidoptera* and Its Contribution to Our Knowledge of Evolution." *Proceedings of the Tenth International Congress of Entomology, 1956* 2 (1958): 831–41. Reprinted in *The Process of Biology: Primary Sources*, edited by J. J. W. Baker and G. E. Allen, 309–22. Reading, MA: Addison-Wesley, 1970.

———. "Darwin's Missing Evidence." *Scientific American* 200, no. 3 (1959): 48–53.

Kimler, W. C. One Hundred Years of Mimicry: History of an Evolutionary Exemplar. PhD diss., Cornell University, 1983a.

———. "Mimicry: Views of Naturalists and Ecologists before the Modern Synthesis." In *Dimensions of Darwinism: Themes and Counterthemes in Twentieth-Century Evolutionary Theory*, edited by M. Grene, 97–127. Cambridge: Cambridge University Press; Paris: Editions de la Maison des Sciences de l'Homme, 1983b.

Kingsland, S. E. *Modeling Nature: Episodes in the History of Population Ecology*. 2nd ed. Chicago: University of Chicago Press, 1995.

———. *The Evolution of American Ecology, 1890–2000*. Baltimore: Johns Hopkins University Press, 2005.

Kinsey, A. C. *The Origin of Higher Categories in* Cynips, Scientific ser., no. 4. Bloomington: Indiana University Publications, 1936.

———. "An Evolutionary Analysis of Insular and Continental Species." *Proceedings of the National Academy of Sciences* 23 (1937): 5–11.

Kipling, R. *Just So Stories*. London: Puffin, 1902.

Kitchin, F. D., W. H. Evans, C. A. Clarke, R. B. McConnell, and P. M. Sheppard. "PTC Taste Response and Thyroid Disease." *British Medical Journal* 1 (1959): 1069–74.

Kleinman, K. "His Own Synthesis: Corn, Edgar Anderson, and Evolutionary Theory in the 1940s." *Journal of the History of Biology* 32 (1999): 293–320.

Kohler, R. E. "Fly Room West: Dobzhansky, *D. Pseudoobscura*, and Scientific Practice." In *The Evolution of Theodosius Dobzhansky*, edited by M. B. Adams, 115–28. Princeton, NJ: Princeton University Press, 1994.

———. *Landscapes and Labscapes: Exploring the Lab-Field Border in Biology*. Chicago: University of Chicago Press, 2002.

Kolchinsky, E. I., ed. *On the Edge: Soviet Science in the First Half of the Twentieth Century*. St. Petersburg, MO: Nauka, 1999.

Kolchinsky, E. I. "Ausgewählte Aspekte der moderne Synthese im russischen Sprachraum zwischen, 1920 und 1940." In *Evolutionsbiologie von Darwin bis heute*, edited by R. Brömer, U. Hossfeld, and N. A. Rupke, 197–210. Berlin: Verlag für Wissenschaft und Bildung, 2000a.

———. "Kurzbiographien Einiger Begründer der Evolutions Synthese in Russland (1920–1940)." In *Evolutionsbiologie von Darwin bis Heute*, edited by R. Brömer, U. Hossfeld, and N. A. Rupke, 211–29. Berlin: Verlag für Wissenschaft und Bildung, 2000b.

Krementsov, N. L. "Dobzhansky and Russian Entomology: The Origin of His Ideas on Species and Speciation." In *The Evolution of Theodosius Dobzhansky*, edited by M. B. Adams, 31–48. Princeton, NJ: Princeton University Press, 1994.

Krimbas, C. B. "Resistance and Acceptance: Tracing Dobzhansky's Influence." In *Genetics of Natural Populations: The Continuing Importance of Theodosius Dobzhansky*, edited by L. Levine, 23–36. New York: Columbia University Press, 1995.

Kuhn, T. S. *The Structure of Scientific Revolutions*. Chicago: University of Chicago Press, 1962.

L'Héritier, P. *Génétique et Évolution: Analyse de Quelques Études Mathématiques sur la Sélection Naturelle.* Paris: Hermann, 1934.

———. "Souvenirs d'un Généticien." *Revue de Synthèse* 102 (1981): 331–50.

L'Héritier, P., Y. Neefs, and G. Teissier. "Aptérisme des Insects et Sélection Naturelle." *Comptes Rendus Hebdomadaire des Séances de l'Académie des Sciences, Paris* 204 (1937): 907–9.

L'Héritier, P., and G. Teissier. "Une Experience de Sélection Naturelle. Courbe d'Élimination du Gène 'Bar' dans une Population de Drosophiles en Équilibre." *Comptes Rendus Hebdomadaires des Séances et Mémoires de la Société de Biologie, Paris* 117 (1934): 1049–51.

Lack, D. "Evolution of the Galapagos Finches." *Nature* 146 (1940): 324–7.

———. "The Galapagos Finches (Geospizinae): A Study in Variation." *Occasional Papers of the California Academy of Sciences*, no. 21 (1945): 1–159.

———. *Darwin's Finches.* Cambridge: Cambridge University Press, 1947.

———. *Evolutionary Theory and Christian Belief: The Unresolved Conflict.* London: Methuen, 1957.

———. *Darwin's Finches: An Essay on the General Biological Theory of Evolution.* Reprint of the 1947 edition with a new preface. New York: Harper, 1961.

———. *Ecological Adaptations for Breeding in Birds.* London: Methuen, 1968.

———. "My Life as an Amateur Ornithologist." *Ibis* 115 (1973): 421–31.

La Follette, M. C. *Creationism, Science, and the Law: The Arkansas Case.* Cambridge, MA: MIT Press, 1983.

Lakatos, I. "Falsification and the Methodology of Scientific Research Programmes." In *Criticism and the Growth of Knowledge*, edited by I. Lakatos and A. Musgrave, 91–196. New York: Cambridge University Press, 1970.

Lamotte, M. "Observations sur la Sélection par les Prédateurs chez *Cepaea nemoralis.*" *Journal de Conchyliologie* 90 (1950): 180–90.

———. "Recherches sur la Structure Génétique des Populations Naturelles de *Cepaea nemoralis* L." *Bulletin Biologique de la France et de la Belgique* 35 (1951): S1–S238.

———. "Le Rôle des Fluctuations Fortuites dans la Diversité des Populations Naturelles de *Cepaea nemoralis* L." *Heredity* 6 (1952): 333–43.

———. "Polymorphism of Natural Populations of *Cepaea nemoralis.*" *Cold Spring Harbor Symposia on Quantitative Biology* 24 (1959): 65–84. Discussion by Rensch, Van Valen, Danserau, Hunt, Dobzhansky, and Lamotte, 84–86.

Lancefield, D. E. "A Genetic Study of Crosses of Two Races or Physiological Species of *Drosophila obscura.*" *Zeitschrift für Induktive Abstammungs und Vererbungslehre* 52 (1929): 287–317.

Laporte, L. F. *George Gaylord Simpson: Paleontologist and Evolutionist.* New York: Columbia University Press, 2000.

Larson, E. J. *Evolution's Workshop: God and Science on the Galápagos Islands.* New York: Basic Books, 2001.

Laubichler, M. D. Review of *Die zweite Darwinsche Revolution*, by T. Junker. *Isis* 97 (2006): 172.

Laudan, L. "A Confutation of Convergent Realism." *Philosophy of Science* 48 (1981): 19–49.

Lawrence, C., and G. Weisz, eds. *Greater than the Parts: Holism in Biomedicine, 1920–1951.* Oxford: Oxford Medical Publications, 1998.

Lederberg, J. "Replica Plating and Indirect Selection of Bacterial Mutants: Isolation of Preadaptive Mutants in Bacteria by Sib Selection." *Genetics* 121 (1989): 395–9. Reprinted in *Perspectives on Genetics*, edited by J. F. Crow and W. F. Dove, 88–92. Madison: University of Wisconsin Press, 2000.

———. Introduction. In *Launching the Antibiotic Era*, edited by C. L. Moberg and Z. A. Cohn. New York: Rockefeller University Press, 1990.

———. "Genetic Recombination in *Escherichia coli*: Disputation at Cold Spring Harbor, 1946–1996." *Genetics* 144 (1996): 439–43. Reprinted in *Perspectives on Genetics*, edited by J. F. Crow and W. F. Dove, 545–50. Madison: University of Wisconsin Press, 2000.

Lederberg, J., and E. M. Lederberg. "Replica Plating and Indirect Selection of Bacterial Mutants." *Journal of Bacteriology* 63 (1952): 399–406.

Lehman, H. "Form of Explanation in Evolutionary Theory." *Theoria* 32 (1966): 14–24.

Leigh, E. G., Jr. "Ronald Fisher and the Development of Evolutionary Theory, I: The Role of Selection." *Oxford Surveys in Evolutionary Biology* 3 (1986): 187–223.

———. "Ronald Fisher and the Development of Evolutionary Theory, II: Influences of New Variation on Evolutionary Process." *Oxford Surveys in Evolutionary Biology* 4 (1987): 212–63.

Lerner, I. M. *The Genetic Basis of Selection*. New York: Wiley, 1958.

———. "The Concept of Natural Selection: A Centennial View." *Proceedings of the American Philosophical Society* 103 (1959): 173–82.

Lerner, I. M., and L. N. Hazen. "Population Genetics of a Poultry Flock under Artificial Selection." *Genetics* 32 (1947): 325–39.

Lessard, S. "Fisher's Fundamental Theorem of Natural Selection Revisited." *Theoretical Population Biology* 52 (1997): 119–36.

Levene, H., L. Ehrman, and R. Richmond. "Theodosius Dobzhansky Up to Now." In *Essays in Evolution and Genetics in Honor of Theodosius Dobzhansky*, edited by M. K. Hecht and W. C. Steere, 1–41. New York: Appleton-Century-Crofts, 1970.

Levine, L., ed. *Genetics of Natural Populations: The Continuing Importance of Theodosius Dobzhansky*. New York: Columbia University Press, 1995.

Lewis, R. W. "Teaching the Theories of Evolution." *American Biology Teacher* 48 (1986): 344–7.

Lewontin, R. C. "Selection In and Out of Populations." In *Ideas in Modern Biology*, edited by J. A. Moore, 299–311. New York: Natural History Press, 1965.

———. "Population Genetics." *Annual Review of Genetics* 1 (1967): 37–70.

———. *Population Biology and Evolution*. Syracuse, NY: Syracuse University Press, 1968a.

———. "The Concept of Evolution." In *International Encyclopedia of the Social Sciences*, vol. 5, edited by D. L. Sills, 202–10. New York: Free Press, 1968b.

———. "Testing the Theory of Natural Selection." *Nature* 236 (1972): 181–2.

———. *The Genetic Basis of Evolutionary Change*. Based on the Jesup Lectures at Columbia University, 1969. New York: Columbia University Press, 1974.

———. "Introduction: The Scientific Work of Th. Dobzhansky." In *Dobzhansky's Genetics of Natural Populations, I–XLIII*, edited by R. C. Lewontin et al., 93–115. New York: Columbia University Press, 1981.

———. "Facts and the Factitious in Natural Sciences." *Critical Inquiry* 18 (1991): 140–153.

———. "Theodosius Dobzhansky: A Theoretician without Tools." In *Genetics of Natural Populations*, edited by L. Levine, 87–101. New York: Columbia University Press, 1995.

———. "Dobzhansky's *Genetics and the Origin of Species:* Is It Still Relevant?" *Genetics*

147 (1997): 351–5. Reprinted in *Perspectives on Genetics*, edited by J. F. Crow and W. F. Dove, 612–6. Madison: University of Wisconsin Press, 2000.

———. "What Do Population Geneticists Know and How Do They Know It?" In *Biology and Epistemology*, edited by R. Creath and J. Maienschein, 191–214. Cambridge: Cambridge University Press, 2000.

Lewontin, R. C., D. Paul, J. Beatty, and C. S. Krimbas. "Interview of R. C. Lewontin." In *Thinking about Evolution*, vol. 2, edited by R. S. Singh et al., 22–61. New York: Cambridge University Press, 2001.

Lindsey, A. W. *The Problems of Evolution*. New York: Macmillan, 1931.

Lipton, P. "Accommodation or Prediction?" *Science* 308 (2005): 1411–2.

Lloyd, E. A. *The Structure and Confirmation of Evolutionary Theory*. Princeton, NJ: Princeton University Press, 1988.

Losee, J. *Theories on the Scrap Heap: Scientists and Philosophers on the Falsification, Rejection, and Replacement of Theories*. Pittsburgh: University of Pittsburgh Press, 2005.

Lustig, A. "Introduction: Biologists on Crusade." In *Darwinian Heresies*, edited by A. Lustig et al., 1–13. Cambridge: Cambridge University Press, 2004.

Lustig, A., R. J. Richards, and M. Ruse, eds. *Darwinian Heresies*. Cambridge: Cambridge University Press, 2004.

MacBride, E. W. *Evolution*. New York: Cape and Smith, 1929.

———. "Mortality amongst Plants and Its Bearing on Natural Selection." *Nature* 125 (1930): 972–3.

Maienschein, J. "Epistemic Styles in German and American Embryology." *Science in Context* 4 (1991): 407–27.

Malécot, G. *Probabilités et Hérédité*. Paris: Presses Universitaires de France, 1966.

———. *The Mathematics of Heredity*. Revised, edited, and translated by D. M. Yermanus. San Francisco: Freeman, 1969.

Manser, E. "The Concept of Evolution." *Philosophy* 40 (1965): 18–34.

Mason, H. L., and J. H. Langenheim. "Natural Selection as an Ecological Concept." *Ecology* 42 (1961): 158–165.

Mather, K. "Polygenic Inheritance and Natural Selection." *Biological Reviews of the Cambridge Philosophical Society* 18 (1943): 32–64.

———. "The Genetical Structure of Populations." In *Evolution*, edited by R. Brown and J. F. Danielli, 66–95. New York: Academic Press, 1953.

———. *Genetical Structure of Populations*. New York: Halsted, 1973.

Mather, K., and L. G. Wigan. "The Selection of Invisible Mutations." *Proceedings of the Royal Society of London* 131 (1942): 50–64.

Matthen, M., and A. Ariew. "Two Ways of Thinking about Fitness and Natural Selection." *Journal of Philosophy* 99 (2002): 55–83.

Mavor, J. W. *General Biology*. 3rd ed. New York: Macmillan, 1947.

Mayr, E. *Systematics and the Origin of Species from the Viewpoint of a Zoologist*. New York: Columbia University Press, 1942.

———. "Some Evidence in Favor of a Recent Date." *Lloydia* 8 (1945): 70–83.

———. "The Bearing of the New Systematics on General Problems: The Nature of Species." *Advances in Genetics* 2 (1948): 205–37.

———. Discussion with B. H. Burmce. *Evolution* 3 (1949a): 369–72.

———. "Speciation and Selection." *Proceedings of the American Philosophical Society* 93 (1949b): 514–9.

———. "Where Are We?" *Cold Spring Harbor Symposia on Quantitative Biology* 24 (1959): 1–14.

———. "Cause and Effect in Biology." *Science* 134 (1961): 1501–6.

———. *Animal Species and Evolution.* Cambridge, MA: Harvard University Press, 1963.

———. *Populations, Species and Evolution.* Cambridge, MA: Harvard University Press, 1970.

———. "The Role of Systematics in the Evolutionary Synthesis." In *The Evolutionary Synthesis*, edited by E. Mayr and W. B. Provine, 123–36. Cambridge, MA: Harvard University Press, 1980a.

———. "How I Became a Darwinian." In *The Evolutionary Synthesis*, edited by E. Mayr and W. B. Provine, 413–23. Cambridge, MA: Harvard University Press, 1980b.

———. "G. G. Simpson." In *The Evolutionary Synthesis*, edited by E. Mayr and W. B. Provine, 452–63. Cambridge, MA: Harvard University Press, 1980c.

———. "Prologue: Some Thoughts on the History of the Evolutionary Synthesis." In *The Evolutionary Synthesis*, edited by E. Mayr and W. B. Provine, 1–48. Cambridge, MA: Harvard University Press, 1980d.

———. *The Growth of Biological Thought: Diversity, Evolution, and Inheritance.* Cambridge, MA: Harvard University Press, 1982.

———. "How Biology Differs from the Physical Sciences." In *Evolution at a Crossroads*, edited by D. J. Depew and B. H. Weber, 43–63. Cambridge, MA: MIT Press, 1985.

———. *Toward a New Philosophy of Biology: Observations of an Evolutionist.* Cambridge, MA: Harvard University Press, 1988.

———. "Controversies in Retrospect." *Oxford Surveys in Evolutionary Biology* 8 (1992a): 1–34.

———. "Haldane's *Causes of Evolution* after 60 Years." *Quarterly Review of Biology* 67 (1992b): 175–86.

———. "Goldschmidt and the Evolutionary Synthesis: A Response." *Journal of the History of Biology* 30 (1997): 31–33.

———. Preface. In *The Evolutionary Synthesis: Perspectives on the Unification of Biology*, edited by E. Mayr and W. B. Provine. Reprint, Cambridge, MA: Harvard University Press, 1998.

———. "Thoughts on the Evolutionary Synthesis in Germany." In *Die Entstehung der synthetische Theorie*, edited by T. Junker and E.-M. Engels, 19–29. Berlin: Verlag der Wissenschaft und Bildung, 1999.

———. *What Makes Biology Unique? Considerations on the Autonomy of a Scientific Discipline.* Cambridge, UK: Cambridge University Press, 2004.

Mayr, E., E. G. Linsley, and L. Usinger. *Methods and Principles of Systematic Zoology.* New York: McGraw-Hill, 1953.

Mayr, E., and W. B. Provine, eds. *The Evolutionary Synthesis: Perspectives on the Unification of Biology.* Cambridge, MA: Harvard University Press, 1980. Reprinted with a new preface by E. Mayr. Cambridge, MA: Harvard University Press, 1998.

Mayr, E., and E. Stresemann. "Polymorphism in the Chat Genus *Oenanthe* (Aves)." *Evolution* 4 (1950): 291–300.

McKerrow, J. C. *Evolution without Natural Selection.* London: Longmans, 1937.

McKusick, V. A., ed. *Medical Genetic Studies of the Amish.* Selected papers assembled and with commentary by V. A. McKusick. Baltimore: Johns Hopkins University Press, 1978.

McOuat, G., and M. P. Winsor. "J. B. S. Haldane's Darwinism in Its Religious Context." *British Journal for the History of Science* 28 (1995): 227–31.

Medawar, P. "Remarks by the Chairman." In *Mathematical Challenges to the Neo-Darwinian Interpretation of Evolution*, edited by P. S. Moorhead and M. M. Kaplan. Philadelphia: Wistar Institute Press, 1967.

Merrell, D. J. *Evolution and Genetics: The Modern Theory of Evolution.* New York: Holt, Rinehart and Winston, 1962.

Mettler, L. E., and T. G. Gregg. *Population Genetics and Evolution.* Englewood Cliffs, NJ: Prentice-Hall, 1969.

Miller, A. I. *Einstein, Picasso: Space, Time, and the Beauty that Causes Havoc.* New York: Basic Books, 2001.

Millstein, R. L. "Random Drift and the Omniscient Viewpoint." *Philosophy of Science* 63 (1996): S10–S18.

———. "Are Random Drift and Natural Selection Conceptually Distinct?" *Biology and Philosophy* 17 (2002): 33–53.

———. "Selection vs. Drift: A Response to Brandon's Reply." *Biology and Philosophy* 20 (2005): 171–5.

———. "Natural Selection as a Population-Level Causal Process." *British Journal for the Philosophy of Science* 57 (2006): 627–53.

———. "Distinguishing Drift and Selection Empirically: 'The Great Snail Debate' of the 1950s." *Journal of the History of Biology* 41 (2007a): 339–67.

———. "Concepts of Drift and Selection in 'The Great Snail Debate' of the 1950s and early 1960s." In *Descended from Darwin: Insights into American Evolutionary Studies, 1900–1970*, edited by J. Cain and M. Ruse, 271–98. Philadelphia: American Philosophical Society, 2009.

Moberg, C. L. "René Dubos: A Harbinger of Microbial Resistance to Antibiotics." *Perspectives in Biology and Medicine* 42 (1999): 559–80.

———. *René Dubos: Friend of the Good Earth.* Washington, DC: American Society for Microbiology Press, 2005.

Moody, P. A. *Introduction to Evolution.* New York: Harper, 1953.

———. *Introduction to Evolution.* 2nd ed. New York: Harper, 1962.

———. *Introduction to Evolution.* 3rd ed. New York: Harper, 1970.

Moore, E. *Heredity: Mainly Human.* London: Chapman and Hall, 1934.

Moore, J. "R. A. Fisher: A Faith Fit for Eugenics." *Studies in History and Philosophy of Biological and Biomedical Sciences* 38 (2007): 110–35.

Moore, J. A. "Science as a Way of Knowing: Evolutionary Biology." *American Zoologist* 24 (1984): 467–534.

Moore, R. *Man, Time, and Fossils: The Story of Evolution.* New York: Knopf, 1953.

Moorhead, P. S., and M. M. Kaplan, eds. *Mathematical Challenges to the Neo-Darwinian Interpretation of Evolution: A Symposium Held at the Wistar Institute of Anatomy and Biology, April 25 and 26, 1966.* Philadelphia: Wistar Institute Press, 1967.

Morgan, T. H. *Scientific Basis of Evolution.* New York: Norton, 1932.

Morris, H. M., ed. *Scientific Creationism.* Public school ed. San Diego: Creation-Life Publishers, 1974.

Morrison, M. "Physical Models and Biological Contexts." *Philosophy of Science* 64 (1997): S315–S324.

———. *Unifying Scientific Theories.* New York: Cambridge University Press, 2000.

———. "Population Genetics and Population Thinking: Mathematics and the Role of the Individual." *Philosophy of Science* 71 (2004): 1189–1200.

Mourant, A. E. "Human Blood Groups and Natural Selection." *Cold Spring Harbor Symposia on Quantitative Biology* 24 (1959): 57–63.

Muller, H. J. "Genetic Variability, Twin Hybrids and Constant Hybrids, in a Case of Balanced Lethal Factors." *Genetics* 3 (1918): 422–99.

———. "Artificial Transmutation of the Gene." *Science* 66 (1927): 84–87.

———. "The Method of Evolution." *Scientific Monthly* 29 (1929): 481–205.

———. "Some Genetic Aspects of Sex." *American Naturalist* 66 (1932): 118–38.

———. "On the Incomplete Dominance of the Normal Allelomorphs of White in *Drosophila*." *Journal of Genetics* 30 (1935): 407–14.

———. "The Present Status of the Mutation Theory." *Current Science* (Bangalore), special no. (1938): 4–45.

———. "Thomas Hunt Morgan, 1866–1945." *Science* 103 (1946): 550–1.

———. "Redintegration of the Symposium on Genetics, Paleontology, and Evolution." In *Genetics, Paleontology and Evolution*, edited by G. L. Jepsen, E. Mayr, and G. G. Simpson, 421–45. Princeton, NJ: Princeton University Press, 1949.

———. *Studies in Genetics: The Selected Papers of H. J. Muller*. Bloomington: Indiana University Press, 1962.

Nagylaki, T. "Gustave Malécot and the Transition from Classical to Modern Population Genetics." *Genetics* 122 (1989): 253–68. Reprinted in *Perspectives on Genetics*, edited by J. F. Crow and W. F. Dove, 103–18. Madison: University of Wisconsin Press, 1989.

Nanay, B. "Can Cumulative Selection Explain Adaptation?" *Philosophy of Science* 72 (2005): 1099–112.

Neel, J. V., and F. M. Salzano. "A Prospectus for Genetic Studies of the American Indian." *Cold Spring Harbor Symposia on Quantitative Biology* 29 (1964): 84–98.

Newman, H. H. *Evolution, Genetics and Eugenics*. 3rd ed. Chicago: University of Chicago Press, 1932.

Norton, B. "Fisher and the Neo-Darwinian Synthesis." In *Human Implications of Scientific Advance: Proceedings of the Fifteenth International Congress of the History of Science; Edinburgh, 10–15 August 1977*, edited by E. G. Forbes, 481–94. Edinburgh, Scotland: Edinburgh University Press, 1978.

———. "Fisher's Entrance into Evolutionary Science: The Role of Eugenics." In *Dimensions of Darwinism*, edited by M. Grene, 19–29. Cambridge: Cambridge University Press; Paris: Editions de la Maison des Sciences de l'Homme,1983.

Norton, H. T. J., and R. C. Punnett. Appendix I. In *Mimicry in Butterflies*, edited by R. C. Punnett, 154–6. Cambridge: Cambridge University Press, 1915.

Numbers, R. L. "Ironic Heresy: How Young-Earth Creationists Came to Embrace Rapid Microevolution by Means of Natural Selection." In *Darwinian Heresies*, edited by A. Lustig, R. J. Richards, and M. Ruse, 84–100. Cambridge: Cambridge University Press, 2004.

Nye, M. J. *Molecular Reality: A Perspective on the Life of Jean Perrin*. New York: American Elsevier, 1972.

Olby, R. "La Théorie Génétique de la Selection Naturelle vue par un Historien." *Revue de Synthèse*, 3rd ser., 103/104 (1981): 251–89.

———. "Huxley's Place in Twentieth-Century Biology." In *Julian Huxley: Biologist and Statesman of Science*, edited by C. K. Waters and A. Van Helden, 53–75. Houston, TX: Rice University Press, 1992.

Orr, H. A. "Dobzhansky, Bateson, and the Genetics of Speciation." *Genetics* 144 (1996):

1331–5. Reprinted in *Perspectives on Genetics*, edited by J. F. Crow and W. F. Dove, 555–9. Madison: University of Wisconsin Press, 2000.

Osborn, H. F. "The Nine Principles of Evolution Revealed by Paleontology." *American Naturalist* 66 (1932): 52–60.

Painter, T. S. "A New Method for the Study of Chromosome Aberrations and the Plotting of Chromosome Maps in *Drosophila melanogaster*." *Genetics* 19 (1934): 175–88.

Park, T. "Sewall Wright: The Chicago Years." *Perspectives in Biology and Medicine* 34 (1991): 497–504.

Parker, G. H. *What Evolution Is.* Cambridge, MA: Harvard University Press, 1925.

Patterson, J. T., and H. J. Muller. "Are 'Progressive' Mutations Produced by X-Rays?" *Genetics* 15 (1930): 495–577.

Patterson, J. T., and W. S. Stone. *Evolution in the Genus* Drosophila. New York: Macmillan, 1952.

Paul, D. B. "The Selection of the 'Survival of the Fittest.'" *Journal of the History of Biology* 21 (1988): 411–24.

Paul, D. B., and C. B. Krimbas. "Nikolai V. Timoféeff-Ressovsky." *Scientific American* 266, no. 2 (1992): 86–92.

Pearl, R. "Requirements of a Proof that Natural Selection Has Altered a Race." *Scientia* 47 (1930): 175–86.

Pfeifer, J. "Why Selection and Drift Might Be Different." *Philosophy of Science* 72 (2005): 1135–45.

Pigliucci, M., and J. Kaplan. *Making Sense of Evolution: The Conceptual Foundations of Evolutionary Biology.* Chicago: University of Chicago Press, 2006.

Pilpel, A. "Statistics Is Not Enough: Revisiting Ronald A. Fisher's Critique (1936) of Mendel's Experimental Results (1866)." *Studies in History and Philosophy of Biological and Biomedical Sciences* 38 (2007): 618–26.

Plough, H. H., and P. T. Ives. "Induction of Mutations by High Temperature in *Drosophila*." *Genetics* 20 (1935): 42–69.

Plutynski, A. "Modeling Evolution in Theory and Practice." *Philosophy of Science* 68, no. 3 (2001): S225–S236.

———. "Explanation in Classical Population Genetics." *Philosophy of Science* 71 (2004): S1201-S1214.

———. "Parsimony and the Fisher-Wright Debate." *Biology and Philosophy* 20 (2005): 697–713.

———. "What Was Fisher's Fundamental Theorem of Natural Selection and What Was It for?" *Studies in History and Philosophy of Biological and Biomedical Sciences* 37 (2006): 59–82.

———. "Evolutionary Biology: Causes, Consequences and Controversies." Review of books by S. Okasha and J. M. Kaplan. *Metascience* 16 (2007a): 437–45.

———. "Drift: A Historical and Conceptual Overview." *Biological Theory* 7, no. 2 (2007b): 156–67.

Popham, E. J. "Experimental Studies of the Biological Significance of Non-cryptic Pigmentation with Special Reference to Insects." *Proceedings of the Zoological Society of London* 117 (1947): 768–83.

Popper, K. *Logik der Forschung: Zur Erkenntnistheorie in der Modernen Naturwissenschaft.* Vienna: J. Springer, [1934] 1935.

———. "The Poverty of Historicism, III." *Economica* 12 (1945): 69–89.

———. *The Poverty of Historicism.* London: Routledge and Kegan Paul, 1957.

———. *The Logic of Scientific Discovery.* London: Hutchinson, 1959.

———. *The Poverty of Historicism.* 2nd ed. London: Routledge and Kegan Paul, 1960.

———. *Conjectures and Refutations.* New York: Basic Books, 1962.

———. *Objective Knowledge: An Evolutionary Approach.* Oxford: Oxford University Press, 1972.

———. "Autobiography of Karl Popper." In *The Philosophy of Karl Popper*, edited by P. A. Schilpp, 3–181. LaSalle, IL: Open Court, 1974.

———. "Natural Selection and the Emergence of Mind." *Dialectica* 32 (1978): 339–55.

Poulton, E. B. "A Hundred Years of Evolution." *Science* 74 (1931): 345–60.

Powell, J. R. "'In the Air': Theodosius Dobzhansky's *Genetics and the Origin of Species.*" *Genetics* 117 (1987): 363–6. Reprinted in *Perspectives on Genetics*, edited by J. F. Crow and W. F. Dove, 31–34. Madison: University of Wisconsin Press, 2000.

———. "The Neo-Darwinian Legacy for Phylogenetics." In *Genetics of Natural Populations*, edited by L. Levine, 71–86. New York: Columbia University Press, 1995.

Price, G. R. "Fisher's 'Fundamental Theorem' Made Clear." *Annals of Human Genetics* 36 (1972): 129–40.

Prout, T. "Four Decades of Inversion Polymorphism and Dobzhansky's Balancing Selection." In *Genetics of Natural Populations*, edited by L. Levine, 49–55. New York: Columbia University Press, 1995.

Provine, W. B. *The Origins of Theoretical Population Genetics.* Chicago: University of Chicago Press, 1971.

———. "The Role of Mathematical Population Geneticists in the Evolutionary Synthesis of the 1930s and 1940s." *Studies in the History of Biology* 2 (1978): 167–92.

———. "Francis B. Sumner and the Evolutionary Synthesis." *Studies in the History of Biology* 3 (1979): 211–40.

———. Introduction (England). In *The Evolutionary Synthesis*, edited by E. Mayr and W. B. Provine, 329–334. Cambridge, MA: Harvard University Press, 1980.

———. "Origins of the Genetics of Natural Populations Series." In *Dobzhansky's Genetics of Natural Populations, I–XLIII*, edited by R. C. Lewontin, J. A. Moore, W. B. Provine, and B. Wallace, 1–83. New York: Columbia University Press, 1981.

———. "The Development of Wright's Theory of Evolution: Systematics, Adaptation, and Drift." In *Dimensions of Darwinism*, edited by M. Grene, 43–70. Cambridge: Cambridge University Press; Paris: Editions de la Maison des Sciences de l'Homme, 1983.

———. "The R. A. Fisher-Sewall Wright Controversy," *Oxford Surveys in Evolutionary Biology* 2 (1985): 197–219. Reprinted in *The Founders of Evolutionary Genetics*, edited by S. Sarkar, 201–29. Dordrecht, Netherlands: Kluwer, 1992.

———. *Sewall Wright and Evolutionary Biology.* Chicago: University of Chicago Press, 1986.

———. "Progress in Evolution and Meaning in Life." In *Evolutionary Progress*, edited by M. Nitecki, 49–74. Chicago: University of Chicago Press, 1988. Reprinted in *Julian Huxley*, edited by C. K. Waters and A. Van Helden, 165–80. Houston, TX: Rice University Press, 1992.

———. "The Origin of Dobzhansky's *Genetics and the Origin of Species.*" In *The Evolution of Theodosius Dobzhansky*, edited by M. B. Adams, 99–114. Princeton, NJ: Princeton University Press, 1994.

———. *The Origins of Theoretical Population Genetics.* 2nd ed. Reprint of the first edition with a new afterword. Chicago: University of Chicago Press, 2001.

....

———. "Ernst Mayr: Genetics and Speciation." *Genetics* 167 (2004): 1041–6.

Punnett, R. C. "Forty Years of Evolutionary Theory." In *Background to Modern Science*, edited by J. Needham and W. Pagel, 189–222. Cambridge: Cambridge University Press, 1938.

Quayle, H. J. "The Development of Resistance to Hydrocyanic Gas in Certain Scale Insects." *Hilgardia* 11 (1938): 183–225.

Reif, W.-E., T. Junker, and U. Hossfeld. "The Synthetic Theory of Evolution: General Problems and the German Contribution to the Synthesis." *Theory in Biosciences* 119 (2000): 41–91.

Reindl, J. "Believers in an Age of Heresy? Oskar Vogt, Nikolai Timofeeff-Ressovsky and Julius Hallervorden at the Kaider Wilhelm Institute for Brain Research." In *Science in the Third Reich*, edited by M. Szöllösi-Janze, 211–42. Oxford: Berg, 2001.

Reinig, W. F. "Die Evolutionsmechanismen, erläutert an den Hummeln." *Zoologischer Anzieger* supplement: *Verhandlungen deutsche Zoologische Gesellschaft, 41,* 12 (1939): 170-206.

Reisman, K., and P. Forber. "Manipulation and the Causes of Evolution." *Philosophy of Science* 72 (2005): 1113–23.

Rensch, B. "The Abandonment of Lamarckian Explanations: The Case of Climatic Parallelism of Animal Characteristics." In *Dimensions of Darwinism*, edited by M. Grene, 31–42. Cambridge: Cambridge University Press; Paris: Editions de la Maison des Sciences de l'Homme Grene, 1983.

Reydon, T. A. C. "On the Nature of the Species Problem and the Four Meanings of 'Species.' " *Studies in History and Philosophy of Biological and Biomedical Sciences* 36 (2005): 135–58.

Rhodes, F. H. T. *The Evolution of Life*. Baltimore: Penguin, 1962.

Richards, O. W., and G. C. Robson. "The Species Problem and Evolution." *Nature* 117 (1926): 345–7, 382–4.

Richards, R. J. *Darwin and the Emergence of Evolutionary Theories of Mind and Behavior*. Chicago: University of Chicago Press, 1987.

Richardson, R. C., and T. C. Kane. "Orthogenesis and Evolution in the 19th Century." In *Evolutionary Progress*, edited by M. Nitecki, 149–68. Chicago: University of Chicago Press, 1988.

Roberts, J. A. F. "Blood Groups and Susceptibility to Disease: A Review." *British Journal of Preventive and Social Medicine* 11 (1957): 107–25.

Robson, G. C. *The Species Problem*. Edinburgh, Scotland: Oliver and Boyd, 1928.

Robson, G. C., and O. W. Richards. *The Variation of Animals in Nature*. London: Longmans, 1936.

Roman, H. "A Diamond in a Desert." *Genetics* 119 (1988): 739–41. Reprinted in *Perspectives on Genetics*, edited by J. F. Crow and W. F. Dove, 58–60. Madison: University of Wisconsin Press, 2000.

Ross, H. H. *A Synthesis of Evolutionary Theory*. Englewood Cliffs, NJ: Prentice-Hall, 1962.

Rudge, D. W. "The Complementary Roles of Observation and Experiment: Theodosius Dobzhansky's Genetics of Natural Populations, IX and XII." *History and Philosophy of the Life Sciences* 22 (2000a): 167–86.

———. "Does Being Wrong Make Kettlewell Wrong for Science Teaching?" *Journal of Biological Education* 35 (2000b): 5–11.

———. "The Role of Photographs and Films in Kettlewell's Popularizations of the Phenomenon of Industrial Melanism." *Science and Education* 12 (2003): 261–87.

. . . .

———. "Did Kettlewell Commit Fraud? Re-examining the Evidence." *Public Understanding of Science* 14 (2005): 249–68.

———. "H. B. D. Kettlewell's Research, 1937–1953: The Influence of E. B. Ford, E. A. Cockayne and P. M. Sheppard." *History and Philosophy of Life Sciences* 28 (2006): 359–88.

Rudwick, M. J. S. "The Inference of Function from Structure in Fossils." *British Journal for the Philosophy of Science* 15 (1964): 27–40.

Ruse, M. "Confirmation and Falsification of Theories of Evolution." *Scientia* 104 (1969): 329–57.

———. "Karl Popper's Philosophy of Biology." *Philosophy of Science* 44 (1977): 638–61. Reprinted in *But Is It Science?*, edited by M. Ruse, 156–76. Buffalo, NY: Prometheus Books, 1988.

———. *Is Science Sexist? And Other Problems in the Biomedical Sciences.* Dordrecht, Netherlands: Reidel, 1981.

———. *Monad to Man: The Concept of Progress in Evolutionary Biology.* Cambridge, MA.: Harvard University Press, 1996a.

———. "Are Pictures Really Necessary? The Case of Sewall Wright's 'Adaptive Landscapes.'" In *Picturing Knowledge: Historical and Philosophical Problems Concerning the Use of Art in Science*, edited by B. S. Baigre, 303–37. Toronto: University of Toronto Press, 1996b.

———. Review of *The Structure of Evolutionary Theory*, by S. J. Gould. *Isis* 95 (2003): 397–8.

———. "Adaptive Landscapes and Dynamic Equilibrium: The Spencerian Contribution to Twentieth-Century American Evolutionary Biology." In *Darwinian Heresies*, edited by A. Lustig, R. J. Richards, and M. Ruse, 131–50. Cambridge: Cambridge University Press, 2004.

Sagan, C. *Cosmos.* New York: Random House, 1980.

Sarkar, S., ed. *The Founders of Evolutionary Genetics: A Centenary Appraisal.* Dordrecht, Netherlands: Kluwer, 1992.

———. "Evolutionary Theory in the 1920s: The Nature of the 'Synthesis.'" *Philosophy of Science* 71 (2004): 1215–26.

Savage, J. M. *Evolution.* New York: Holt, Rinehart and Winston, 1963.

Scerri, E. R., and J. Worrall. "Prediction and the Periodic Table." *Studies in History and Philosophy of Science* 32A (2001): 407–52.

Schindewolf, O. *Paläontologie, Entwicklungslehre und Genetik: Kritik und Synthese.* Berlin: Borntraeger, 1936.

———. *Grundfragen der Paläontologie.* Stuttgart, Germany: Schweizerbart, 1950. Reprinted in English with the title *Basic Questions in Paleontology: Geologic Time, Organic Evolution, and Biological Systematics.* Translated by J. Schaefer Chicago: University of Chicago Press, 1993.

Schmalhausen, I. I. *Factors of Evolution: The Theory of Stabilizing Selection.* Edited by T. Dobzhansky. Translated by I. Dordick. Chicago: University of Chicago Press, 1949. First published in Russian with the title *Faktory Evolyutsii: Teoria stabiliziryushevo Otbora.* Moscow: Nauka, 1946.

Schmitt, S. "L'oeuvre de Richard Goldschmidt: Une Tentative de Synthèse de la Génétique, de la Biologie et de la Théorie de l'Evolution autour du concept d'Homéose." *Revue d'Histoire des Sciences* 53 (2000): 381–99.

....

Scott, E. C. *Evolution vs. Creationism: An Introduction.* Westport, CT: Greenwood Press, 2004.

Scriven, M. "The Age of the Universe." *British Journal for the Philosophy of Science* 5 (1954): 181–90.

———. "Explanation and Prediction in Evolutionary Theory." *Science* 130 (1959): 477–82.

Shanahan, T. *The Evolution of Darwinism: Selection, Adaptation, and Progress in Evolutionary Biology.* New York: Cambridge University Press, 2004.

Shapere, D. "Biology and the Unity of Science." *Journal of the History of Biology* 22 (1969): 3–18.

Sheppard, P. M. "Fluctuations in the Selective Value of Certain Phenotypes in the Polymorphic Land Snail *Cepaea nemoralis* L." *Heredity* 5 (1951): 125–34.

———. "Evolution in Bisexually Reproducing Organisms." In *Evolution as a Process*, edited by J. S. Huxley et al., 201–19 London: Allen and Unwin, 1954.

———. "Genetic Variability and Polymorphism: A Synthesis." Published with discussion by M. Lamotte and P. M. Sheppard. *Cold Spring Harbor Symposia on Quantitative Biology* 20 (1955): 271–5.

———. *Natural Selection and Heredity.* London: Hutchinson, 1958.

———. "Blood Groups and Natural Selection." *British Medical Journal* 15 (1959a): 134–9.

———. "The Evolution of Mimicry: A Problem in Ecology and Genetics." *Cold Spring Harbor Symposia on Quantitative Biology* 24 (1959b): 131–40.

———. *Natural Selection and Heredity.* 3rd ed. London: Hutchinson, 1967.

Shull, A. F. *Evolution.* New York: McGraw-Hill, 1936.

———. *Evolution.* 2nd ed. New York: McGraw-Hill, 1951.

Shull, A. F., G. R. Larue, and A. G. Ruthven. *Principles of Animal Biology.* 5th ed. New York: McGraw-Hill, 1941.

Simon, C. W. "Insecticide Resistance versus Antimicrobial Resistance: Biological Issues in Historical Perspective." *Gesnerus* 60 (2003): 235–59.

Simpson, G. G. *Tempo and Mode in Evolution.* New York: Columbia University Press, 1944.

———. "Essay-Review of Recent Works on Evolutionary Theory by Rensch, Zimmerman, and Schindewolf." *Evolution* 3 (1949a): 178–84.

———. *The Meaning of Evolution.* New Haven, CT: Yale University Press, 1949b.

———. "Trends in Research and the *Journal of Paleontology*." *Journal of Paleontology* 24 (1950): 498–9.

———. *The Major Features of Evolution.* New York: Columbia University Press, 1953a.

———. "The Baldwin Effect." *Evolution* 7 (1953b): 110–7.

———. *This View of Life: The World of an Evolutionist.* New York: Harcourt, Brace and World, 1964.

———. *The Meaning of Evolution.* Rev. ed. New Haven: Yale University Press, 1967.

———. "Uniformitarianism: An Inquiry into Principle, Theory, and Method in Geohistory and Biohistory." In *Essays in Evolution and Genetics*, edited by M. K. Hecht and W. C. Steere, 43–96. New York: Appleton-Century-Crofts, 1970.

Simpson, G. G., and W. S. Beck. *Life: An Introduction to Biology.* 2nd ed. New York: Harcourt, Brace and World, 1965.

Singh, R. S., C. B. Krimbas, D. B. Paul, and J. Beatty, eds. *Thinking about Evolution: Historical, Philosophical, and Political Perspectives.* Cambridge: Cambridge University Press, 2001.

Skipper, R. A., Jr. "Selection and the Extent of Explanatory Unification." *Philosophy of Science* 66 (1999): S196–S209.

———. "The R. A. Fisher-Sewall Wright Controversy in Philosophical Focus: Theory Evaluation in Population Genetics." PhD diss., University of Maryland, College Park, 2000.

———. "The Persistence of the R. A. Fisher-Sewall Wright Controversy." *Biology and Philosophy* 17 (2002): 341–67.

———. "The Heuristic Role of Sewall Wright's Adaptive Landscape Diagram." *Philosophy of Science* 71 (2004): 1176–88.

Skipper, R. A., Jr., and R. L. Millstein. "Thinking about Evolutionary Mechanisms: Natural Selection." *Studies in History and Philosophy of Biological and Biomedical Sciences* 36 (2005): 327–47.

Slatis, H. M. "The Study of Normal Variation in Man, II: Polymorphism and Pleiotropy." *Cold Spring Harbor Symposia on Quantitative Biology* 29 (1964): 61–68.

Sloan, P. R. "Ernst Mayr on the History of Biology." *Journal of the History of Biology* 18 (1985): 145–53.

Slobodkin, L. B. "Toward a Predictive Theory of Evolution." In *Population Biology and Evolution*, edited by R. C. Lewontin, 187–205. Syracuse, NY: Syracuse University Press, 1968.

Smith, J. M. "The Status of Neo-Darwinism." In *Towards a Theoretical Biology*, vol. 2, edited by C. H. Waddington, 82–89. Chicago: Aldine, 1969.

———. Interview. In *A Passion for Science*, edited by L. Wolpert and A. Richards, 124–37. Oxford: Oxford University Press, 1988.

———. "Genetics, Evolution and Haldane." *Quarterly Review of Biology* 67 (1992): 187–9.

Smocovitis, V. B. "Botany and the Evolutionary Synthesis: The Life and Work of G. Ledyard Stebbins, Jr." PhD diss., Cornell University, 1988.

———. *Unifying Biology: The Evolutionary Synthesis and Evolutionary Biology*. Princeton, NJ: Princeton University Press, 1996.

———. "G. Ledyard Stebbins, Jr. and the Evolutionary Synthesis, 1924–1950." *American Journal of Botany* 84 (1997): 1625–37.

———. "Keeping Up with Dobzhansky: G. Ledyard Stebbins, Jr., Plant Evolution, and the Evolutionary Synthesis." *History and Philosophy of the Life Sciences* 28 (2006): 9–48.

———. "The 'Plant *Drosophila*': E. B. Babcock, the Genus *Crepis*, and the Evolution of a Genetics Research Program at Berkeley, 1915–1947." *Historical Studies in the Natural Sciences* (forthcoming).

Snyder, L. H. *Principles of Heredity*. 2nd ed. Boston: Heath, 1940.

Sober, E. *The Nature of Selection*. Cambridge, MA: MIT Press, 1984a.

———. *Conceptual Issues in Evolutionary Biology*. Cambridge, MA: MIT Press, 1984b.

———. *The Philosophy of Biology*. Boulder, CO: Westview Press, 1993.

———. "Evolution and Optimality: Feathers, Bowling Balls, and the Thesis of Adaptationism." *1995–1996 Philosophic Exchange*, no. 26 (1996): 41–57.

———. ed. *Conceptual Issues in Evolutionary Biology*. 3rd ed. Cambridge, MA: MIT Press, 2006.

Stadler, L. J. "Mutations in Barley Induced by X-rays and Radium." *Science* 68 (1928): 186–7.

Stamos, D. N. "Popper, Falsifiability, and Evolutionary Biology." *Biology and Philosophy* 11 (1996): 161–91.

Stebbins, G. L., Jr. "Evolutionary Factors and the Fossil Evidence." Letter to E. H. Colbert, 8 June 1944. *Bulletin of the Committee on Common Problems in Genetics, Paleontology and Systematics*, no. 3, 1944. Reprinted in *Exploring the Borderlands: Documents of the Committee on Common Problems in Genetics, Paleontology and Systematics, 1943–1944.* Vol. 94, pt. 2, *Transactions of the American Philosophical Society*, edited by J. Cain, 67–70. Philadelphia: American Philosophical Society, 2004.

———. "Evidence for Abnormally Slow Rates of Evolution, with Particular Reference to the Higher Plants and the Genus *Drosophila*." *Lloydia* 8 (1945): 84–102.

———. "Reality and Efficacy of Selection in Plants." *Proceedings of the American Philosophical Society* 93 (1949): 501–13.

———. *Variation and Evolution in Plants*. Based on the Jessup Lectures at Columbia University in 1946. New York: Columbia University Press, 1950.

———. "The Dynamics of Evolutionary Change." In *Lectures in Biological Sciences*, edited by J. I. Townsend, 39–62. Knoxville: University of Tennessee Press, 1959a.

———. "The Synthetic Approach to Problems of Organic Evolution." *Cold Spring Harbor Symposia on Quantitative Biology* 24 (1959b): 305–11.

———. *Processes of Organic Evolution*. Englewood Cliffs, NJ: Prentice-Hall, 1966.

———. *The Basis of Progressive Evolution*. Chapel Hill: University of North Carolina Press, 1969.

———. "Variation and Evolution in Plants: Progress During the Past Twenty Years." In *Essays in Evolution and Genetics*, edited by M. K. Hecht and W. C. Steere, 173–208. New York: Appleton-Century-Crofts, 1970.

———. "Botany and the Synthetic Theory of Evolution." In *The Evolutionary Synthesis*, edited by E. Mayr and W. B. Provine, 139–52. Cambridge, MA: Harvard University Press, 1980.

Stephens, C. "Selection, Drift, and the 'Forces' of Evolution." *Philosophy of Science* 71 (2004): 550–70.

Stone, L., and P. F. Lurquin. *A Genetic and Cultural Odyssey: The Life and Work of L. Luca Cavalli-Sforza*. New York: Columbia University Press, 2005.

Struthers, D. "ABO Groups of Infants and Children Dying in the West of Scotland (1949–1951)." *British Journal of Preventive and Social Medicine* 5 (1951): 223–228.

Sukatschew, W. "Einige Experimentelle Untersuchungen über den Kampf ums Dasein Zwischen Biotypen Derselben Art." *Zeitschrift für Induktive Abstammungs und Vererbungslehre* 47 (1928): 57–74.

Sumner, F. B. "Genetic, Distributional, and Evolutionary Studies of the Subspecies of Deer-Mice (*Peromyscus*)." *Bibliographia Genetica* 99 (1932): 1–106.

———. "Evidence for the Protective Value of Changeable Coloration in Fishes." *American Naturalist* 69 (1935): 245–66.

———. "Where Does Adaptation Come in?" *American Naturalist* 76 (1942): 433–44.

———. *The Life History of an American Naturalist*. Lancaster, PA: Jacques Cattell Press, 1945.

Tennyson, A. *In Memoriam*. London: Moxon, 1850.

Thompson, P. "'Organization,' 'Population,' and Mayr's Rejection of Essentialism in Biology." In *Aristotle and Contemporary Science*, vol. 2, edited by D. Mentzou, J. Hattiangadi, and F. M. Johnson, 173–83. New York: Lang, 2000.

Thomson, J. A. *The System of Animate Nature: The Gifford Lectures Delivered in the University of St. Andrews in the Years 1915 and 1916.* London: Williams and Norgate, 1920.

Thomson, J. A., and P. Geddes. *Life: Outlines of General Biology.* Vol. 2. New York: Harper, 1931.

Timofeeff-Ressovsky, N. W. "Zur Analyse des Polymorphismus bei *Adalia punctata.*" *Biologische Zentralblatt* 60 (1940): 130–7.

Toulmin, S. *Foresight and Understanding.* New York: Harper Torchbooks, 1961.

———. "The Evolutionary Development of Natural Science." *American Scientist* 55 (1966): 456–71.

Truesdell, C. "The Rational Mechanics of Flexible or Elastic Bodies: Introduction to *Leonhardi Euleri Opera Omnia,*" 2nd ser., vols. 10/11. Zürich, Switzerland: Füssli, 1960.

Turner, J. R. G. "Fisher's Evolutionary Faith and the Challenge of Mimicry." *Oxford Surveys in Evolutionary Biology* 2 (1985): 59–196.

———. "Random Genetic Drift, R. A. Fisher, and the Oxford School of Ecological Genetics." In *The Probabilistic Revolution,* vol. 2, edited by L. Krüger et al., 313–54. Cambridge, MA: MIT Press, 1987.

———. "Kettlewell, Henry Bernard Davis." In *Dictionary of Scientific Biography,* vol. 17, no. 2, edited by F. L. Holmes, S469–S471. New York: Charles Scribner's Sons, 1990.

Villee, C. A. *Biology: The Human Approach.* 2nd ed. Philadelphia: Saunders, 1954.

Vucinich, A. "Russia: Biological Sciences." In *The Comparative Reception of Darwinism,* edited by T. F. Glick, 227–68. Chicago: University of Chicago Press, 1974.

Waddington, C. H. *An Introduction to Modern Genetics.* New York: Macmillan, 1939.

———. "Epigenetics and Evolution." *Evolution, Symposia of the Society for Experimental Biology* 7 (1953): 186–99.

———. *The Strategy of the Gene: A Discussion of Some Aspects of Theoretical Biology.* London: Allen and Unwin, 1957.

———. "Evolutionary Adaptation." In *Evolution after Darwin,* vol. 1, edited by S. Tax, 381–402. Chicago: University of Chicago Press, 1960.

———. *The Nature of Life.* London: Allen and Unwin, 1961.

———. "Paradigm for an Evolutionary Process." In *Towards a Theoretical Biology,* vol. 2, edited by C. H. Waddington, 106–24. Chicago: Aldine, 1969..

Waddington, C. H., et al. *Science and Ethics: An Essay.* London: Allen and Unwin, 1942.

Wade, M. J. "Sewall Wright: Gene Interaction and the Shifting Balance Theory." *Oxford Surveys in Evolutionary Biology* 8 (1992): 35–62.

Wade, M. J., and C. J. Goodnight. "Wright's Shifting Balance Theory: An Experimental Study." *Science* 253 (1991): 1015–8.

Wallace, B. *Chromosomes, Giant Molecules, and Evolution.* New York: Norton, 1966.

Wallace, B., and A. M. Srb. *Adaptation.* Englewood Cliffs, NJ: Prentice-Hall, 1961.

Wallace, C. *A Study of Evolution.* Glasgow, Scotland: Blackie, 1969.

Walsh, D. M., T. Lewens, and A. Ariew. "The Trials of Life: Natural Selection and Random Drift." *Philosophy of Science* 69 (2002): 452–73.

Waters, C. K., and A. Van Helden, eds. *Julian Huxley: Biologist and Statesman of Science.* Proceedings of a conference held at Rice University in 1987. Houston, TX: Rice University Press, 1993.

Watson, D. M. S. *Paleontology and Modern Biology.* Based on lectures delivered in 1937. New Haven, CT: Yale University Press, 1951.

———. "Is 'Macroevolution' Reality?" *Transactions of the New York Academy of Sciences*, 2nd ser., 14 (1952): 302–3.

Watson, D. M. S., et al. "Discussion on the Present State of the Theory of Natural Selection." *Proceedings of the Royal Society of London* B121 (1936): 43–73.

Weatherwax, P. *Plant Biology*. 2nd ed. Philadelphia: Saunders, 1947.

Weber, B. H., and D. J. Depew, eds. *Evolution and Learning: The Baldwin Effect Reconsidered*. Cambridge, MA.: MIT Press, 2003.

Weisz, P. B. *Elements of Biology*. 2nd ed. New York: McGraw-Hill, 1965.

———. *The Science of Zoology*. New York: McGraw-Hill, 1966.

———. *Elements of Zoology*. New York: McGraw-Hill, 1968.

Wells, H. G., J. S. Huxley, and G. P. Wells. *The Science of Life*. Garden City, NY: Doubleday, 1931.

Wells, J. "Second Thoughts about Peppered Moths." In *Darwinism, Design, and Public Education*, edited by J. A. Campbell and S. C. Meyer, 187–92. East Lansing: Michigan State University Press, 2003.

White, M. J. D. *Animal Cytology and Evolution*. Cambridge: Cambridge University Press, 1945.

Willermet, C. M., and B. Hill. "Fuzzy Set Theory and Its Implications for Speciation Models." In *Conceptual Issues in Modern Human Origins Research*, edited by G. A. Clark and C. M. Willermet, 77–88. New York: Aldine de Gruyter, 1997.

Williams, G. C. *Adaptation and Natural Selection: A Critique of Some Current Evolutionary Thought*. Princeton, NJ: Princeton University Press, 1966.

Williams, M. B. "Deducing the Consequences of Evolution: A Mathematical Model." *Journal of Theoretical Biology* 29 (1970): 343–85.

———. "Falsifiable Predictions of Evolutionary Theory." *Philosophy of Science* 40 (1973): 518–37.

———. "Similarities and Differences between Evolutionary Theory and the Theories of Physics." *PSA 1980* 2: 385–96. In *Proceedings of the 1980 Biennial Meeting of the Philosophy of Science Association*, vol. 2, edited by P. Asquith and R. Giere. East Lansing, MI: Philosophy of Science Association, 1981.

———. "The Importance of Prediction Testing in Evolutionary Biology." *Erkenntnis* 17 (1982): 291–306.

———. "The Scientific Status of Evolutionary Theory." *American Biology Teacher* 47 (1985): 205–10.

Wilson, E. O. *Sociobiology: The New Synthesis*. Cambridge, MA: Harvard University Press, 1975.

Winsor, M. P. "Species, Demes, and the Omega Taxonomy: Gilmour and *The New Systematics*." *Biology and Philosophy* 15 (2000): 349–88.

———. "Ernst Mayr, 1904–2005." *Isis* 96 (2005): 415–8.

Wolters, G., and J. G. Lennox, eds. *Concepts, Theories, and Rationality in the Biological Sciences*. In collaboration with P. McLaughlin. Konstanz, Germany: Universitätsverlag; Pittsburgh: University of Pittsburgh Press, 1995.

Wright, S. "Evolution in Mendelian Populations." *Genetics* 16 (1931): 97–159.

———. "The Roles of Mutation, Inbreeding, Crossbreeding, and Selection in Evolution." *Proceedings of the Sixth International Congress of Genetics* 1 (1932): 356–66. Reprinted in *Evolution*, edited by G. E. Brosseau Jr., 68–78. Dubuque, IA: Brown, 1967.

———. "The Statistical Consequences of Mendelian Heredity in Relation to Speciation." In *The New Systematics*, edited by J. Huxley, 161–83. Oxford: Clarendon Press, 1940.

———.Review of *The Material Basis of Evolution*, by R. Goldschmidt. *Scientific Monthly* 53 (1941): 165–70.

———. "On the Role of Directed and Random Changes in Gene Frequency in the Genetics of Populations." *Evolution* 2 (1948): 279–94.

———. "Fisher and Ford on 'The Sewall Wright Effect.'" *American Scientist* 39 (1951): 452–8, 479.

———. "Classification of the Factors of Evolution." *Cold Spring Harbor Symposia on Quantitative Biology* 20 (1955): 16–24.

———. "Physiological Genetics, Ecology of Populations, and Natural Selection." In *Evolution after Darwin*, vol. 1, edited by S. Tax, 429–75. Chicago: University of Chicago Press, 1960.

———. "The Foundations of Population Genetics." In *Heritage from Mendel*, edited by R. A. Brink, 245–63. Madison: University of Wisconsin Press, 1967.

———. *Evolution: Scientific Papers*. Edited and with introductory materials by W. B. Provine. Chicago: University of Chicago Press, 1986.

Wright, S., and T. Dobzhansky. "Genetics of Natural Populations, XII: Experimental Reproduction of Some of the Changes Caused by Natural Selection in Certain Populations of *Drosophila pseudoobscura*." *Genetics* 31 (1946): 125–56.

Wright, S., and W. E. Kerr. "Experimental Studies of the Distribution of Gene Frequencies in Very Small Populations of *Drosophila melanogaster*, II: Bar." *Evolution* 8 (1954): 225–40.

Yoon, C. K. "Ernst Mayr, Pioneer in Tracing Geography's Role in the Origin of Species, Dies at 100." *New York Times*, 5 February 2005, sec. A.

Zachos, F. "Karl Popper und die Biologie: Zur Falsifizierbarkeit der Evolutionshypothese und der Selektionstheorie." In *Die Entstehung biologischer Disziplinen, II*, edited by U. Hossfeld and T. Junker, 171–94. Berlin: Verlag für Wissenschaft und Bildung, 2002.

Zahar, E. "Why Did Einstein's Programme Supersede Lorentz's?" *British Journal for the Philosophy of Science* 24 (1973): 95–123, 233–62.

INDEX

....

....